# Steel Corrosion in Concrete

## Fundamentals and Civil Engineering Practice

**Arnon Bentur**

*National Building Research Institute–Faculty of Civil Engineering, Technion, Israel Institute of Technology, Haifa, Israel*

**Sidney Diamond**

*School of Civil Engineering, Purdue University, W. Lafayette, Indiana, USA*

**Neal S. Berke**

*W.R. Grace & Co. - Conn., Grace Construction Products, Cambridge, Massachusetts, USA*

CRC Press is an imprint of the
Taylor & Francis Group, an informa business
A SPON PRESS BOOK

CRC Press
Taylor & Francis Group
6000 Broken Sound Parkway NW, Suite 300
Boca Raton, FL 33487-2742

First issued in paperback 2019



Typeset in 10/12pt Palatino by Saxon Graphics Ltd, Derby

ISBN-13: 978-0-419-22530-0 (hbk)
ISBN-13: 978-0-367-86371-5 (pbk)



A catalogue record for this book is available from the British Library

**Visit the Taylor & Francis Web site at**
**http://www.taylorandfrancis.com**

**and the CRC Press Web site at**
**http://www.crcpress.com**

# Contents

CHAPTER 1

# Introduction

Reinforced concrete is a relatively new construction material which has been developed and applied extensively only in the 20th century. It has always been stated that the combination of concrete and reinforcing steel is an optimal one not only because of the mechanical performance but also from the point of view of long-term performance. Concrete is a durable material, much more than steel, and the encasement of steel in it provides the steel with a protective environment and allows it to function effectively as a reinforcement. Theoretically, this combination should be highly durable, as the concrete cover over the steel provides a chemical and physical protection barrier to the steel, and can potentially eliminate steel corrosion problems which occur readily in bare steel structures.

The tendency of bare steel to corrode is well known, and we are all aware of the sophisticated technologies developed to protect it, such as paint systems and active electrochemical methods (e.g. cathodic protection). Even with these protective means tedious maintenance procedures can not be avoided and their costs can be quite high. Encasement of steel in concrete can theoretically provide durable and maintenance-free construction material. Experience with many structures has demonstrated that this can be indeed the case, except perhaps in some severe environmental conditions.

In spite of the theory and favourable performance record in many structures, corrosion of steel in concrete has become in the past three decades a considerable durability problem in mild as well as in severe climatic conditions. Whereas in the past we were mainly concerned with the performance of the concrete itself, e.g. resistance of concrete to sulphate attack typical of marine structures, it seems that, at present, the most common durability problem is corrosion of steel in concrete. Performance in marine structures is an example of the 'turning around' in durability concerns: although many of the text books and specifications give considerable attention to the concrete itself in these environmental conditions and specify the use of sulphate-resistant cement, the major durability problem in this environment is the corrosion of the steel in concrete. This corrosion usually occurs before any noticeable sulphate attack, and it is the 'bottle-neck' in the durability performance of reinforced concrete structures in such environment.

The increased incidence of durability problems involving steel corrosion in reinforced concrete structures is the result of several changes which have taken place in concrete technology and in the environmental conditions in which concrete is being increasingly used. Some of these will be outlined here:

- Since the performance of concrete is specified in terms of 28-days' strength, the modern cements are designed to achieve higher strength levels and obtain most of their strength potential within 28 days (i.e. higher $C_3S$ content and higher fineness), leading to the use of concretes with higher w/c ratio and less reserve for post 28 days' hydration. These two factors reduce the effective long-term protection of the concrete cover.
- The need to optimize the structure has led to a tendency to reduce the concrete cover depth, which when coupled with less attention to proper workmanship can result in effective reduction in the protection offered by the concrete cover.
- Many environments have become more severe compared with those for which the structure was originally designed. De-icing by chloride salts is carried out more frequently to meet the needs of the heavier winter traffic in cold climates, leading to an effectively higher chloride ingress in bridge decks and parking garages (chloride is a major agent inducing corrosion of steel). Another example is the increasing corrosive nature of the environment in industrially polluted zones.
- Increase use of architectural concrete in which the reinforced concrete component is not protected with an additional mortar rendering.
- Increased use of concrete in marine structures and in urban structures along the sea shore where the concentration of chloride in the environment is higher than usual.

The higher incidence of corrosion of steel in concrete is well documented. The damage is usually observed first as rust stains and minute cracking over the concrete surface (Fig. 1.1) and frequently the cracks run in straight lines parallel to the underlying reinforcement (Fig. 1.2). This type of damage is the result of the increase in volume associated with the formation of the corrosion products (i.e. rust). Such damage can be observed usually in 'critical locations' such as parts of the structure where humidity is more readily maintained, or at the base of columns in contact with the soil where there is a greater tendency for accumulation of salts due to capillary rise. If repair means are not taken at this early stage, the corrosion of the steel will proceed further, causing severe damage through delamination and spalling (Figs 1.2 and 1.3), as well as exposure of the steel and reduction of its cross-section to an extent which may

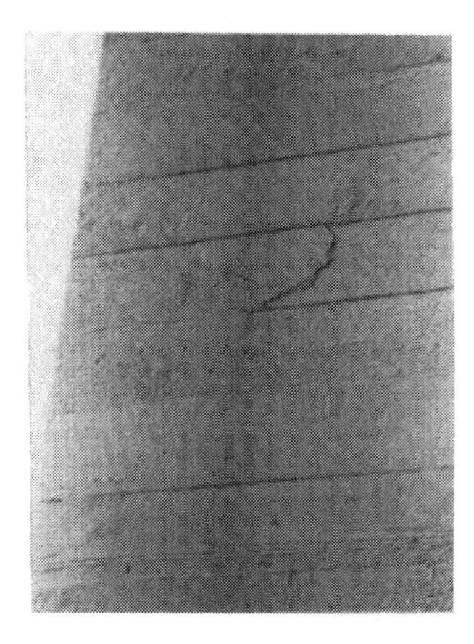

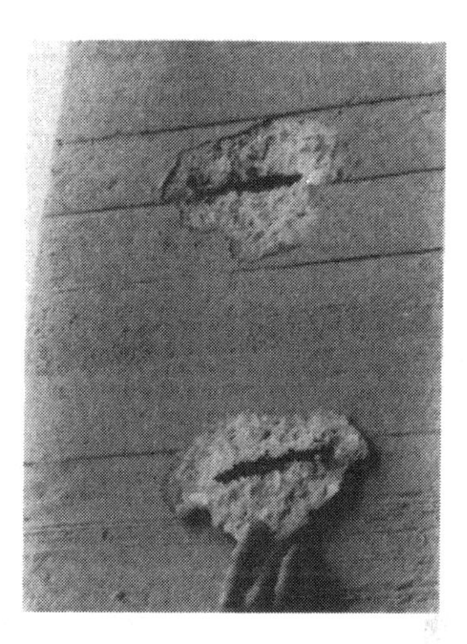

**Fig. 1.1.** (A) Local corrosion damage on exposed concrete showing up as 'swelling' and local cracking. (B) The damaged concrete can be readily removed, exposing beneath it the corroded reinforcing bar. (Reproduced courtesy of C. Jaegerman.)

**Fig. 1.2.** Corrosion damage in a column which shows up as longitudinal cracks parallel to the main reinforcement located at the corner of the column (bottom part of the column). The damaged concrete can be readily removed, exposing beneath it the corroded reinforcing bar at the corner of the column (upper part of the column). (Reproduced courtesy of C. Jaegerman.)

**Fig. 1.3.** Corrosion damage in densely reinforced wall, resulting in delamination due to simultaneous corrosion of the reinforcing bars. (Reproduced courtesy of C. Jaegerman.)

become a safety hazard. Structures where such damage can be quite critical are bridges (Fig. 1.4) and marine structures (Fig. 1.5).

The economic implications of such damages are quite large. For example, in the case of bridge decks in the USA, the yearly cost of repairs is estimated to be between $50 to $200 millions [1.1]. The repair operations themselves are quite complex and require special treatments of the cracked zones, and in most instances the life expectancy of the repair is limited. Also, additional indirect costs have to be considered due to inactivation of the structure and temporary loss of its economic benefits.

Thus, on the one hand it is stated that reinforced concrete structures can be inherently durable, while in practice we are confronting many cases of damage due to corrosion of steel in concrete. There is no apparent contradiction between the two; reinforced concrete structures can be constructed to be durable to corrosion of steel problems. We have a reasonably good grasp of the basic scientific concepts, and the problem is usually in the application of these concepts to engineering practices and design. The object of this book is to present the science of the phenomenon of corrosion of steel in concrete, and to apply it to practice by discussing the means and design methods to achieve a durable structure. Special attention is given to critical review of available standards and specifications, emphasizing the context within which they can be successfully applied and the cases where their applicability is limited. One may refer to it as 'from the science to the engineering of elimination of corrosion of steel in concrete'.

**Fig. 1.4.** Corrosion in a bridge structure. (A) General view showing damage in the column and the beam. (B) Damage in the column showing up as cracks and peeling of concrete due mainly to corrosion of the main longitudal reinforcing bars – the cracks run mainly parallel to the main reinforcement. (C) Damage in the beam showing up as cracks, rust stains and exposure of the reinforcing bars; the damage is of a 'linear type' parallel to the stirrups, suggesting that they are the ones to corrode. (Reproduced courtesy of C. Jaegerman.)

**Fig. 1.5.** Damage in piles in a marine structure, showing heavy delamination and corrosion of the bars of the densely reinforced structure. (Reproduced courtesy of C. Jaegerman.)

## REFERENCES

1.1. Menzies, T.R., National Cost of Damage from Highway De-icing. In *Corrosion Forms and Control for Infrastructure*, ed. V. Chaker. ASTM STP 1137, American Society of Testing and Materials, Philadelphia, 1992, pp. 30–45.

CHAPTER 2

# Mechanisms Of Steel Corrosion

Corrosion of steel in any environment is a process that involves progressive removal of atoms of iron (Fe) from the steel being corroded. The iron is removed by an electrochemical reaction and is dissolved in the surrounding water solution, appearing as ferrous ($Fe^{2+}$) ions. In steel embedded in concrete, the dissolution takes place in the limited volume of water solution present in the pores of the concrete surrounding the steel. As a result of this dissolution process, the steel loses mass, i.e. its cross-section becomes smaller.

If the steel is part of a reinforced concrete structure under load, the stress carried by the remaining cross-section will necessarily increase. In extreme cases, such an increase may impose a safety risk or may even lead to failure. This is one of the obvious risks associated with steel corrosion in concrete.

However, there are other perhaps less obvious risks. The ferrous ions dissolved in the concrete pore water solutions usually react with hydroxide ($OH^-$) ions and dissolved oxygen ($O_2$) molecules to form one of several varieties of rust, which is a solid by-product of the corrosion reaction. The rust is usually deposited in the restricted space in the concrete around the steel. Its formation within this restricted space sets up expansive stresses which may crack the concrete cover. This in turn may result in progressive deterioration of the concrete, especially when freezing and thawing or other environmental influences prevail.

## 2.1. CORROSION REACTIONS

The corrosion of steel in concrete is an electrochemical process in which both chemical reactions and flow of electrical current are involved. The chemical and electrical processes are coupled. In order to evaluate conditions that might lead to steel corrosion in concrete, one needs to be familiar with at least the basic concepts of both the chemical reactions and the electrical processes involved. Such familiarity is also necessary if one is to design a reinforced concrete structure so as to minimize the risk of steel corrosion difficulties occurring in service.

The basic concepts of the corrosion processes are not overly complex. We shall attempt to describe them in simple terms, emphasizing only those aspects that are important in understanding steel corrosion in concrete. The reader is referred to standard references, such as those by Uhlig [2.1], Scully [2.2] and Fontana [2.3].

The corrosion process actually involves two separate, but coupled, chemical reactions that take place simultaneously at two different sites on the steel surface. As indicated in Fig. 2.1, an electric current must flow in a closed loop between the two sites for the reactions to proceed. The overall process is quite similar to what takes place in an ordinary dry cell battery by means of which electricity is generated. Such batteries are technically described as 'galvanic cells', so the corrosion process is sometimes called 'galvanic corrosion'.

In corrosion, the two electrochemical reactions are known as 'anodic' and 'cathodic' reactions, respectively, and the areas on which they occur in the steel are called 'anodic' and 'cathodic' areas or simply anodes and cathodes, as indicated in Fig. 2.1. The two reactions are as follows:

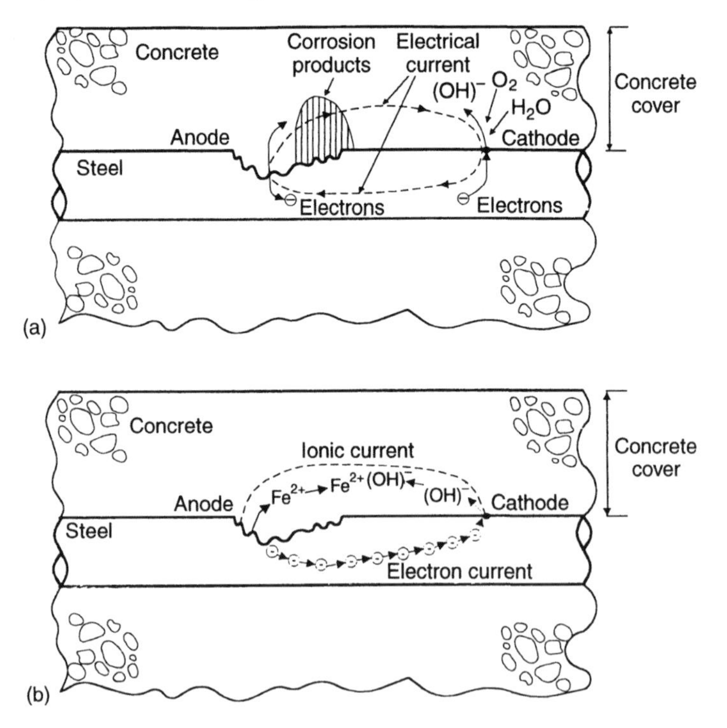


**Fig. 2.1.** Corrosion processes on the surface of steel. (A) Reactions at anodic and cathodic sites and electric current loop. (B) Flow of electrical charge in the electric current loop during the corrosion process. The direction of the current in Fig. 2.1(A) is represented by the accepted sign convention: positive direction is the direction of flow of positive charge. In Fig. 2.1(A) the flow of the positive charge is presented. In Fig. 2.1(B) the flow of the actual charge is shown.

| | | | | |
|---|---|---|---|---|
| Anodic reaction | | | | |
| $2\ Fe^0$ – | 4 electrons | → | $2Fe^{2+}$ | (2.la) |
| metallic atoms at the steel surface | | | ions dissolved in solution | |
| Cathodic reaction | | | | |
| $O^{2-} + 2H_2O$ + | 4 electrons | → | $4(OH^-)$ | (2.lb) |
| dissolved oxygen molecules | | | ions dissolved in solution | |

The flow of electrons from the anodic areas where they are produced by the anodic reaction (2.la) to the cathodic areas where they are used in the cathodic reaction (2.lb), and its counter-current ionic flow in the external concrete pore solution constitute the corrosion current. Such flow is extremely important, since its interruption will cause the corrosion process to come to a halt.

The actual loss of metal involved in the corrosion processes takes place at the anodic sites, and is indicated by reaction (2.la). The iron atoms are ionized to ferrous ($Fe^{2+}$) ions which dissolve in the water solution around the steel. The electrons are deposited on the steel surface and raise its electrical potential. The electrons then flow through or along the steel to a lower potential (cathodic) site, as indicated in Fig. 2.l(B). Here, reaction (2.lb) takes place, and the electrons combine with dissolved oxygen molecules and water to form hydroxide ($OH^-$) ions. In order for the corrosion process to continue, the number of electrons accepted at the cathodic site in reaction (2.lb) needs to be equal to the number of electrons donated at the anodic site (reaction 2.la). Thus, for every dissolved oxygen molecule that reacts at the cathodic site, two iron atoms must be ionized and dissolved at the anodic site. The actual metal removal process described in reaction (2.la) will continue only if there is a cathodic reaction which will act as a sink for the electrons produced at the anodic site. Therefore, if oxygen and water are not available at the cathodic sites, the corrosion process will be terminated.

The electrochemical character of the corrosion process is demonstrated by the flow of current in the closed loop of Fig. 2.l(B). The electrical current within the steel bar is the result of electron flow from the anode to the cathode. An external current also flows through the solution through the pores of the concrete surrounding the steel, carried by the movement of charged ions. This external current consists of hydroxide ions (negatively charged) moving from the cathode to the anode, and also ferrous ions (positively charged) moving from the anode to the cathode. It is important to realize that the 'water' in the concrete pores is actually a dilute solution of alkali and calcium hydroxides; it is this fluid that serves as a vehicle for ionic flow. If the pores are dried out, or if the structure of the concrete is so dense that the pores are not very well interconnected, the flow of ions through the pores becomes difficult. Under such circumstances the ionic current will flow only with difficulty, and the corrosion process will slow down or stop.

The concepts described in the previous paragraph may be restated in purely electrical terms. If the electrical resistance in the concrete (i.e. the resistance to flow of ions) through which the ionic current must pass is high, the rate of flow of the current carried by the ions will be low, and the anodic and cathodic reactions will proceed only slowly. Thus the rate of corrosion will be low. The moving ferrous ($Fe^{2+}$) and hydroxide ($OH^-$) ions flow toward one another. When they meet, they react to form ferrous hydroxide, $Fe(OH)_2$. This compound will react further with additional hydroxide ions, and sometimes also oxygen, to generate the insoluble product that we see as rust.

There are two primary types of rust, red rust which is $Fe_2O_3$, and black rust, $Fe_3O_4$. It is important to appreciate that neither type necessarily deposits at the anodic site. The rust is only a chemical by-product of the corrosion process, and it often accumulates at places other than where the actual corrosion of metal occurs. The actual corrosion process is really the process of converting metal iron atoms to ferrous ions dissolved in the surrounding aqueous solution.

In analysing the practical aspects of steel corrosion in concrete, two distinct aspects need to be addressed: (i) whether corrosion can occur at all; and (ii) if it occurs, at what rate will it proceed.

To assess whether it is likely for corrosion to occur at all, one must understand the conditions that give rise to the formation of anodic and cathodic sites and that initiate a corrosion reaction. If it is known that these conditions will never be present, corrosion will be of no practical concern.

If the conditions for corrosion to occur are present and anodic and cathodic reactions have started, what is the expected rate of these processes? In practical terms, how much of the steel cross-section will be lost per year, and what rate of accumulation of corrosion products may be expected? If it can be expected that the rates will always be low, corrosion again may not pose a practical problem, despite the fact that the corrosion process has begun and that it might be expected to continue indefinitely.

## 2.2. LIKELIHOOD OF CORROSION OCCURRING

Anodic and cathodic areas can be generated on the surface of steel or other metals whenever there are compositional variations from place to place, where local differences in applied stress occur, or where certain variations in local environmental conditions occur. Such differences cause some areas to be more chemically 'active' than others. These active areas may become anodes and corrode, i.e. lose mass. The less active sites will tend to act as cathodes.

Variations in chemical activity from place to place are associated with corresponding differences in electrical potential; it is these electrical potential differences which are the actual driving forces for the corrosion reactions. Large electrical potential differences can result when two dissimilar metals are in contact, with one metal serving as the anode and the other as the cathode. If a wet or humid environment is present, rapid corrosion can occur. However, corrosion can also occur due to potential differences between different areas of the same piece of metal, when compositional or other variations occur.

Ordinary reinforcing steel is heterogeneous on a microscopic scale, being an iron carbon alloy containing two different kinds of crystals, the α-iron crystal usually called ferrite, and an iron carbide compound ($Fe_3C$) called cementite, as indicated schematically in Fig. 2.2. The ferrite phase always tends to develop a more active potential than the cementite phase.

Potential differences can also occur in different areas of a metal made up of only a single type of crystal, even when the composition is uniform throughout. The atoms at or near the grain boundaries within the metal tend to be significantly more active than corresponding atoms within the bulk of the grains, and the grain boundary regions tend to become anodic. If cells are set up under such circumstances, they tend to be only of microscopic dimensions, and to be located adjacent to each other. A given site may be permanently anodic or permanently cathodic, but this is not always the case. Sometimes anodic and cathodic areas shift continuously over the surface of the metal.

In any event, many metal surfaces may be viewed as consisting of clusters of local corrosion cells, i.e. combinations of adjacent potentially anodic and cathodic sites. As long as the metal remains dry, local currents will not flow and corrosion will not take place. However, when the metal is exposed to water or ionic solutions, currents will flow and the local

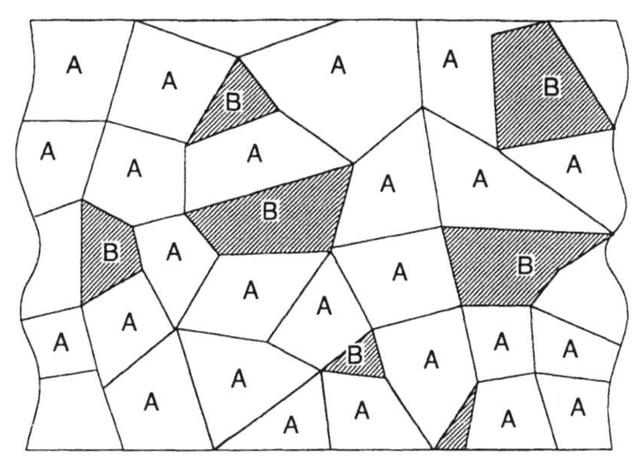


Fig. 2.2. Schematic description of the microstructure of an alloy metal, showing that it is composed of phases of different composition (phases A and B) and individual grains of the same composition. The darker lines between the grains represent grain boundaries.

cells will become active, marking the onset of corrosion. In processes of this nature the corrosion generally will take place more or less uniformly over the metal surface. Thus, this kind of corrosion is often termed 'general' corrosion.

Stresses due to cold working of metals can lead to inhomogeneity effects similar to those produced by grain boundaries. The metal within the strained zone is more active, and will act as the anode and corrode; the unstrained metal will function as the cathode. Since cold working is usually confined to small areas, this type of corrosion tends to be much more localized than the general corrosion mentioned previously.

Differences in actual stresses being applied to different portions of a single metal bar may also give rise to differences in activity and in electrical potential; again the more highly stressed regions tend to become anodic and corrode.

As mentioned previously, corrosion can also occur because of differences in the local environment surrounding different areas of the same piece of metal; this is true even though the metal itself may be quite homogeneous. Such corrosion takes place most commonly when local differences in the concentration of dissolved oxygen occur. If certain zones along the steel are more highly aerated than others, a possibility of corrosion will exist. The aerated zones in which the concentrations of dissolved oxygen are high tend to become cathodic, and the less well-aerated zones, anodic. The corrosion induced is thus of the localized type, and is generally called 'aeration cell corrosion'.

Aeration cell corrosion is one kind of a general class of corrosion types known collectively as 'concentration cell corrosion'. This may be induced whenever there are differences in the local concentrations of any dissolved species in the vicinity of different areas of the steel. Even differences in moisture contents in different parts of the concrete cover surrounding the reinforcing steel can induce concentration cell corrosion.

## 2.3. CORROSION RATES

In the previous section the likelihood of the occurrence of corrosion was discussed, specifically in terms of the possibilities for development of electrical potential differences between potentially anodic and potentially cathodic areas on the same steel. However, the existence of such cells does not necessarily imply that the corrosion process will be of practical concern. The actual rate of corrosion may be so slow that the processes are of little practical significance. Corrosion rates are sensitive to a number of additional factors not yet discussed. If the rate of corrosion can be controlled and kept low, corrosion may occur without presenting a serious practical problem.

One important factor controlling the rate of corrosion is the availability of dissolved oxygen surrounding the cathodic areas. Oxygen is consumed in the cathodic reaction (2.lb) and, if its supply in the solution surrounding the cathodic areas of the metal is not continuously replenished, the corrosion reactions may be retarded.

One way that this can occur is if the surface of the steel is surrounded by a protective layer (for example, concrete cover) which slows down or prevents the diffusion of dissolved or of gaseous oxygen from the surrounding environment. This is shown diagrammatically in Fig. 2.3. In such cases the rate of corrosion becomes 'diffusion controlled', i.e. regulated by the rate of diffusion of oxygen through the concrete cover. The onset of the process of regulation of the rate of corrosion by slow diffusion of oxygen produces a significant reduction in the potential difference between the anodic and cathodic areas. Such an effect is called a 'polarizing effect', and the process is called 'polarization'. Limited diffusion of oxygen is actually only one of several different types of possible causes of polarizing effects that may occur in practice, and it is referred to as 'concentration polarization'.

The bottle-neck in the corrosion process for steel in concrete may often turn out to be the limitation on the flow of ionic (not electronic) current through the pores of the concrete surrounding the steel. If the rate of flow of the charge-carrying ionic species is slow, the corrosion reactions can proceed only at a slow rate. This occurs when the electrical resistance of the concrete surrounding the embedded steel between the anode and the cathode is high, as shown in Fig. 2.3. Thus, measurement of the electrical resistance of the concrete surrounding the embedded steel can sometimes serve as an indication of how fast the corrosion reactions can proceed.

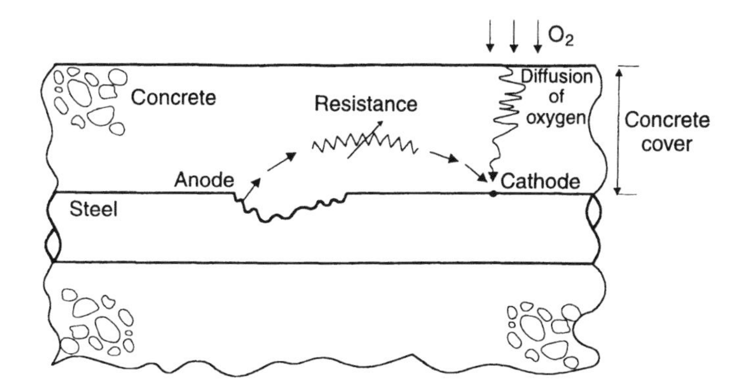


Fig. 2.3. Schematic description of two factors which may slow down the rates of corrosion of steel in concrete. (i) Diffusion of oxygen into cathodic site; in dense and water-saturated concrete, the diffusivity of oxygen is reduced. (ii) Electrical resistance to flow of ionic current: in dense and dry concrete the resistance is high.

Another type of process often limits the rate of corrosion. Some metals, including steel, can react with oxygen to form thin (~10 nm (0.01 μm) thick) layers of insoluble metal oxide on their surfaces. If this film remains stable in contact with the aqueous solution, the metal can be rendered electrochemically passive. In effect, the film isolates the metal from the aqueous solution, and slows down the rate of anionic dissociation to very low values. In fact, as long as a passivating film remains effective, the corrosion rates can be so low that for practical purposes the metal can be considered as non-corroding. This situation is shown in Fig. 2.4(A).

When this condition occurs, the metal is usually referred to as being 'passivated', and the film is called a 'passivating film'. This passivation process should not be confused with concentration polarization, a quite different process discussed earlier. Formation of a passivating film is usually responsible for the corrosion resistance exhibited by stainless steel. The iron in such steel is alloyed with chromium, nickel, and other elements which enhance the formation and stability of the passivating film.

Effective passivation can also be obtained with conventional steel, if it is in contact with aqueous solution in which the concentration of

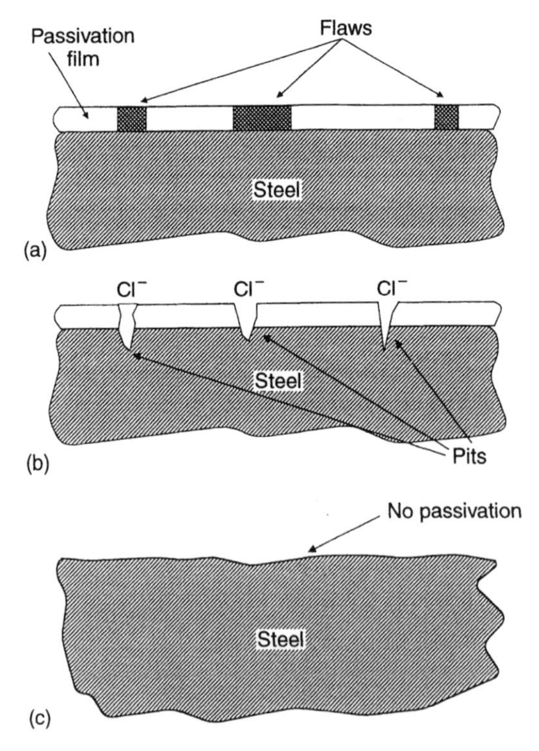


**Fig. 2.4.** Schematic description of the formation and stability of the passivating film. (A) Coherent film, practically preventing corrosion. (B) Formation of pits due to faults induced during the breakdown of the passivating film. (C) Depassivation of the whole passivation film in carbonated concrete where the pH is low.

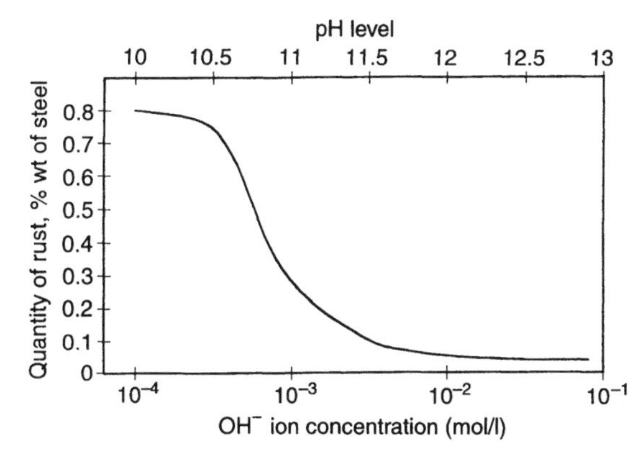


**Fig. 2.5.** Corrosion of steel in water solutions as a function of $OH^-$ ion concentration and pH level; the reduced corrosion at pH 12.5 is the result of formation of effective passivating film. (Adapted from Ref. 2.4)

hydroxide ($OH^-$) ions is high enough. This is indicated in Fig. 2.5, which shows the results of a series of experiments in which the extent of steel corrosion (as estimated by the weight of rust formed) is plotted against the $OH^-$ ion concentration. The level of hydroxide ion concentration required to maintain passivation is not a constant value, but varies with the presence of other ions, especially chloride ions. The concentration of the $OH^-$ ions in the water solution in the pores of the concrete in contact with the steel is usually sufficiently high to maintain this passivation film. Extensive discussion of this issue will be provided in Chapter 4.

In view of the significance of the passivation film that can form in the steel embedded in concrete, it is worthwhile to provide additional details of the reactions involved in its formation and the nature of the conditions that may lead to its destabilization. These are carbonation of the concrete (that leads to reduction in the $OH^-$ ion concentration in the pore solution) and the presence of chloride ($Cl^-$) ions that may diffuse to the pore solution surrounding the steel.

The oxide films that normally develop on steel are either ferrous ($Fe^{2+}$) or ferric ($Fe^{3+}$) in nature. Both are chemically stable in concrete in the absence of carbonation or chloride. However, ferric oxide is the most stable, especially in the presence of chloride; over time the ferrous oxide is converted to the more stable ferric oxide. The latter is chemically referred to as $\gamma$-FeOOH. The conversion process is never totally completed; its effect can be detected as part of the very small passive corrosion rate that continues to be exhibited by passivated steel, as will be discussed in section 6.1.

The development of the film is shown schematically in Fig. 2.4(A) and proceeds according to the following equations:

$$Fe \rightarrow Fe^{2+} + 2e^- \quad (2.2)$$

$$Fe^{2+} + 2OH^- \rightarrow Fe(OH)_2 \quad (2.3)$$

$$Fe(OH)_2 + O_2 \rightarrow \gamma\text{-}FeOOH + H_2O \quad (2.4)$$

Neither ferrous nor ferric oxide films are protective when the $OH^-$ ion concentration is sufficiently low (pH less than 11.5; see later in this section for definition of pH). At lower levels of $OH^-$ ion content, the oxides form, but do not adhere to the steel surface and therefore are not protective. This may occur when the concrete around the steel becomes carbonated, resulting in a reduction in the $OH^-$ ion concentration in the pore solution (for extensive discussion, see section 4.2.1).

As noted above, films that are mostly in the ferric oxide form are more resistant to chlorides than mostly ferrous oxide films. The chlorides react with the ferrous oxide to form a soluble complex which dissolves in the surrounding solution and does not provide protection:

$$Fe(OH)_2 + Cl^- \rightarrow [FeCl]_x \quad (2.5)$$

Thus, if chloride ions are present in the concrete pore solution, they will induce corrosion if they come into contact with the reinforcing bar at a location where the ferrous oxide has not been converted. This is shown in Fig. 2.4(B). The effect is to induce pitting corrosion. This process of local pitting corrosion competes with the final stage in the passivation process, the conversion of ferrous oxide film to ferric oxide, and will only proceed when the chloride content is high compared with the hydroxide ion contents. The high $OH^-$ ion concentration is necessary to ensure stability to the ferric oxide film. Furthermore, dissolved oxygen favours the normal passivation route, by primarily promoting the final step of conversion of ferrous oxide to ferric oxide. Therefore, pitting corrosion will tend to occur preferentially in regions of low dissolved oxygen content. This is the underlying cause of the importance of the chloride-to-hydroxide ratios controlling the onset of pitting corrosion as discussed in section 4.2.2.

Because of the importance of this passivation mechanism in protecting reinforcing steel in concrete, and the importance of the solution composition in determining whether it can be maintained, it is appropriate that we review here certain necessary elements of water solution chemistry. These involve the conventional scales used to express concentrations of ions in solution, and the special scales used specifically to express the concentrations of hydroxide ions and of hydrogen ions.

Solution concentrations are usually expressed in units of molarity or moles per litre; that is, the number of gram-molecular weights of the dissolved substance per litre. Sometimes the concentrations of the specific ions are expressed in units of 'normality', i.e. the number of gram-equivalent weights per litre. The gram equivalent weight may be the same as the gram molecular weight, if we are concerned with substances that dissolve to produce ions each of which has only a single charge – for example hydrochloric acid, HCl, which produces equivalent amounts of singly charged $H^+$ and $Cl^-$ ions. The gram-molecular weight of this substance is 36.5, which is the sum of the gram atomic weights of 1 for hydrogen and 35.5 for chlorine. A solution of 36.5 grams/litre concentration can be conventionally stated as being of 1 molar (1 M) concentration in either HCl or in each of its constituent ions.

In contrast, sulphuric acid ($H_2SO_4$) dissolves to produce two singly charged $H^+$ ions and one doubly charged sulphate ($SO_4^{2-}$) ion per molecule. The gram-molecular weight of sulphuric acid is 98, the sum of 2 for the two hydrogen atoms and 96 for the combination of one sulphur and four oxygen atoms which make up the sulphate ion.

As before, a solution of 1 gram-molecular weight per litre, in this case 98 grams per litre, is spoken of as being a 1 M solution of the compound sulphuric acid, $H_2SO_4$.

The concentration of the sulphate, or $SO_4^{2-}$ ions is also 1 M, since there are exactly as many of these as there were sulphuric acid molecules. However, there are twice as many $H^+$ ions, and so the solution concentration with respect to hydrogen ions is not 1 M, but 2 M. For many people unfamiliar with chemical manipulations, this situation provides a source of confusion.

The concept of the gram-equivalent weight avoids this awkward situation. The gram-equivalent weight of an ion is its gram ionic weight divided by the number of charges it carries. Thus, the gram-equivalent weight of the sulphate ion is its gram ionic weight of 96 divided by 2, or 48. A 1 M solution of 98 grams of sulphuric acid per litre, producing 96 grams of dissolved $SO_4^{2-}$ ions, contains 2 equivalents of sulphate ions rather than 1. The solution thus has a sulphate ion concentration of 2 gram-equivalent weights per litre, i.e. is a solution of 2 N concentration in sulphate.

The solution above also contains 2 grams of dissolved $H^+$ ions. The gram-equivalent weight of hydrogen is its gram ionic weight of 1, divided by its charge, also 1. Thus, the specified solution has 2 gram-equivalent weights of hydrogen ions per litre, i.e. is 2 N with respect to hydrogen ions as well as with respect to sulphate ions.

In general, when we express concentrations in normalities based on gram-equivalent weights of the ions, the same concentration number is obtained for the positively and for the negatively charged ions resulting from dissolution of a neutral molecule. This convention offers advantages also when considering the stoichiometry of reactions in solution. However, in the present context we are concerned primarily with concentrations of singly charged ions: $OH^-$ ions, $H^+$ ions, and $Cl^-$ ions. Since all of these are singly charged, the molar concentration and the normality will be identical for a given solution for each of them.

Nonetheless, when considering concentrations of $OH^-$ and of $H^+$ ions, a new complication occurs. The range of concentration values of these ions in practical solutions is so great that a linear scale of concentration is awkward and a logarithmic scale is ordinarily used. This is expressed in the shorthand notation as the so-called 'pH' value.

The pH value (or simply, the pH) of a given solution is defined as the negative logarithm, to the base 10, of the $H^+$ ion concentration (more properly, chemical activity). Molar concentrations are conventionally indicated by square brackets [ ]; thus, the $H^+$ ion concentration is conventionally symbolized as $[H^+]$. The pH is thus defined as:

$$pH = -\log [H^+] \tag{2.6}$$

For water solutions there is a fixed and invariant relationship between the concentration of the $H^+$ ions and that of the $OH^-$ ions, in that the product of the two must always equal $K_w$, the so-called 'ion product constant' of water:

$$K_w = [H^+]\,[OH^-] \tag{2.7}$$

The ion product constant varies with temperature, but for 'ordinary' temperatures it is approximately $10^{-14}$. From eqns (2.6) and (2.7) it follows that there is a relationship between pH and the $OH^-$ ion concentration through $K_w$:

$$pH = 14 - \log [OH^-] \tag{2.8}$$

In pure water, the only source of the $H^+$ and $OH^-$ ions is the water itself, by dissociation of a very small percentage of the water molecules in the reaction $H_2O \rightarrow H^+ + OH^-$. The two concentrations are equal and are each $10^{-7}$ molar. By either eqn. (2.6) or eqn. (2.8), the pH is 7. Thus, 'neutral' solutions have a pH of 7. Acidic solutions, those in which $H^+$ ions predominate, have pH values less than 7, and alkaline or basic solutions, in which $OH^-$ ions predominate, have pH values greater than 7. A solution that is 1 molar in $OH^-$ ions has a pH of 14; this is not an upper limit, since more concentrated solutions may readily be produced with alkaline substances.

To summarize the issue of the effect of pH on passivation, several points are highlighted again. The effectiveness of the passivating film on a steel surface is a function of the pH of the water solution around the steel. If the pH is generally greater than about 11.5, and certain ions, such as $Cl^-$ are absent, a passivating film is normally maintained and the corrosion rate should be very low. The effectiveness of the passivating film is impaired when the pH is reduced below this level, or when $Cl^-$ ions are present in the solution. At sufficiently high concentrations, chloride ions can destabilize the passivating films, even at pH values in excess of 13, as illustrated in Fig. 2.4(B). This destabilization of the passivation film occurs at discrete locations and results in pitting corrosion. This is different to the situation where the whole passivation film is destabilized when the pH is reduced, and results in general corrosion as shown in Fig. 2.4(C).

## REFERENCES

2.1. Uhlig, H.H., *Corrosion and Corrosion Control*. John Wiley and Sons, Inc., 1971.
2.2. Scully, J.C., *The Fundamentals of Corrosion*. Pergamon Press, 1975.
2.3. Fontana, M.G., *Corrosion Engineering*. McGraw-Hill, 1986.
2.4. Shalon, R. & Raphael, M., Influence of Sea Water on Corrosion of Reinforcement. *J. Am. Concrete Inst.*, **30**(12) (1959), 1251–68.

CHAPTER 3

# Relationships between Corrosion and the Structure and Properties of Concrete

## 3.1. STRUCTURE OF CONCRETE AND OF CEMENT PASTE

The corrosion of steel in reinforced concrete is considerably different to the corrosion of steel exposed to the atmosphere, since the concrete cover around the reinforcing steel changes its chemical environment significantly.

The primary function of the concrete cover with respect to corrosion is to serve as a protective coating. Its effectiveness is a function of both its chemical composition and its physical structure. Important aspects of its physical structure include its total porosity, its pore size distribution and the degree to which the coarser pores are interconnected. We will therefore find it useful to review some of the characteristics of concrete before attempting to discuss the details of the corrosion of steel in concrete.

In this discussion we will assume that the reader is generally familiar with the structure and properties of concrete, and will confine our discussion to those features especially relevant to steel corrosion. For more complete information on concrete structure and properties the reader is referred to such standard texts on concrete as those of Neville [3.1], Mindess and Young [3.2] and Mehta and Monteiro [3.3].

Concrete is a composite material composed of coarse and fine aggregates embedded in a matrix of hardened hydrated Portland cement. The aggregate is usually impermeable and more or less inert chemically. On the other hand, the hardened Portland cement ('cement paste') is porous and contains pore solutions which are chemically active. The nature of the cement paste and its pore structure and pore solution composition will be the dominant factors controlling the effectiveness of the concrete cover.

The cement paste structure develops as a result of chemical reaction between the Portland cement grains and the mix water. Concrete sets at a relatively early stage of hydration, after which the external boundaries of the mass are more or less fixed. However, the volume of the hydrated products that continue to form as a result of the reactions is considerably greater than the volume of the cement grains from which they are derived. In consequence, the continued hydration tends to have a 'space-filling' effect, as shown in highly schematic form in Fig. 3.1.

This effect is rather different for pastes of low and of high water/cement (w/c) ratios, respectively. In both cases the porosity of the cement paste continues to decrease as hydration proceeds. However, it is customary, after Powers [3.4] to view the pores as belonging to two different types: larger so-called 'capillary' pores which are subdivided but not filled up by hydration products; and a much finer class of pores called 'gel' pores, that are developed within the particles of the hydration products, rather than occurring between them. In low w/c concretes, after hydration has continued for some reasonable time, the total porosity will be relatively small. The pore space that exists will occur mostly as ultra-fine gel pores, within the hydration products (Fig. 3.1). The paste structure is much denser and much less permeable than the structure that is produced in high w/c ratio concretes. In concretes batched at a high w/c ratio not only is the total pore volume higher, but a considerable number of interconnected coarse capillary pores will remain even after very long hydration.

As a consequence of these differences, low w/c ratio concrete is intrinsically less permeable than high w/c ratio concrete and, incidentally, is considerably stronger as well. In consequence, many specifications place an upper limit on the w/c ratio to be used for reinforced concrete that is to be exposed to aggressive environmental conditions which could lead to the corrosion of the reinforcing steel.

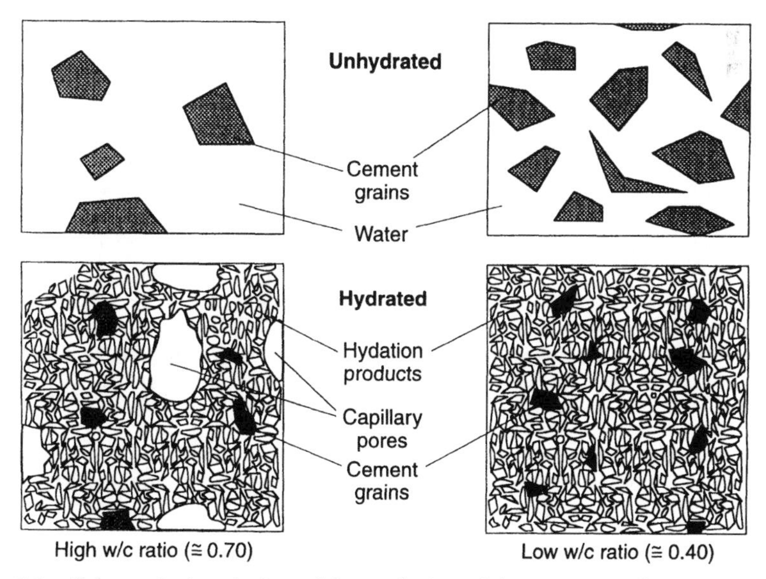


**Fig. 3.1.** Schematic description of the evolution of the structure of cement pastes of low and high w/c ratio.

Unfortunately, batching concrete at a low w/c ratio is not sufficient by itself to prevent corrosion. It is also necessary that the concrete mix composition be rich enough in cement so that a sufficient volume of cement paste is produced to fill the spaces between the aggregate grains. Furthermore, the workability of the mix needs to be such that the concrete can be effectively compacted. As a result, a minimum cement content is often also specified, and it is sometimes specifically indicated that this provision is aimed at achieving high compacted concrete density rather than high strength.

Proper field practice is also essential to secure a concrete cover of proper quality. Without adequate compaction and curing, even properly designed concrete will often not be as impermeable as expected on the basis of its mix composition.

## 3.2. CHEMISTRY OF CONCRETE

The hydration products of Portland cement consist predominantly of two kinds: relatively large crystals of calcium hydroxide [$Ca(OH)_2$] and very small (colloidal) particles of a rather indefinite hydrated calcium silicate compound often indicated in abbreviated form as C-S-H gel. The average composition of the C-S-H gel in most concretes is roughly $3CaO.2SiO_2.2H_2O$. Small amounts of several different kinds of hydrated calcium aluminate sulphates and other products may also occur.

All Portland cements contain at least a few tenths of a percent of alkalis, expressed in analytical terms as $K_2O$ and $Na_2O$, but actually present either as alkali sulphates or alkali-calcium sulphates, or in solid solution within certain compounds of the cement. Much of the alkali content, especially that present as alkali sulphate, is readily soluble in the mix water.

When concrete sets, the residual mix water is enclosed in the pores of the newly solidified material. As curing proceeds, some of this water is used up to form hydrated compounds. Depending on the circumstances of exposure and of the concrete permeability, it may be replaced by additional water from the outside.

The 'water' in the pores is never pure water, however, but always contains dissolved ions derived from the cement. Thus, we usually talk about 'pore solutions' rather than 'pore water'. After several days of hydration, the pore solutions of most concretes usually contain alkali and hydroxide ions in relatively high concentration, along with much smaller concentration of calcium and sulphate ions, and traces of certain others. The alkali ions are potassium ($K^+$) and sodium ($Na^+$) ions; with most modern cements, the former predominates.

The $OH^-$ ion concentration is ordinarily quite high, being usually between 0.5 and 1.0 molar, even when the concrete pores are filled with solution, i.e. 'saturated'. If the concrete partially dries out, concentrations can be higher. Values for pH corresponding to these high concentrations of hydroxide are a approximately 13.

Since the steel embedded in the concrete is in contact with these pore solutions of high alkalinity, a tight and stable passivating film is formed and the steel is ordinarily protected against corrosion indefinitely.

The exact value of the pH found in concrete pore solutions is influenced by several factors, chief among which are the alkali content of the cement used and the richness in cement of the concrete. Other factors such as the w/c ratio, the extent to which alkalis have leached out of the concrete in structures immersed in water, and the extent to which alkalis have reacted with aggregate components to form stable insoluble reaction products may also influence the pH value to some degree. A factor of great importance specifically with respect to the cover, i.e. the outermost layer of the concrete, is the extent to which carbonation has occurred. This will be discussed subsequently.

As indicated previously, passivating films are generally formed and maintained in concrete, and little corrosion of embedded steel occurs. However, sometimes ions other than those previously discussed may be found in the concrete pore solutions. Such ions may be derived from chemical admixtures added to the mix water, contamination of the aggregate, or as the result of penetration of ions into hardened concrete exposed to natural waters or other external solutions.

The ion of greatest concern is the chloride ion, $Cl^-$. Chloride and a few other ions of similar character have the property of effectively destabilizing the passivating film, if present in sufficiently high concentration. This effect is actually one of the two principal causes of problems of steel corrosion in concrete.

Specific sources of chloride ions in concrete include accelerating admixtures that contain calcium chloride, salt-contaminated aggregate, sea water, salt spray, or deliberately applied de-icing salt that comes in contact with concrete.

## REFERENCES

3.1. Neville, A.M., *Properties of Concrete*. 3rd ed., Pitman, 1981.
3.2. Mindess, S. & Young, J.F., *Concrete*. Prentice-Hall, 1981.
3.3. Mehta, P.K. & Monteiro, P.J.M., *Concrete: Structure, Properties, and Materials*. Prentice-Hall, 1993.
3.4. Powers, T.C., The Physical Structure and Engineering Properties of Concrete. *Res. and Dev. Bull.* No. 90, Portland Cement Association, Skokie, IL., 1958.

CHAPTER 4

# Corrosion of Steel in Concrete

## 4.1. INTRODUCTION

The risk of corrosion of reinforcing steel should be minimal in a well-designed reinforced concrete structure containing a sufficient depth of good quality concrete cover which has been properly placed and compacted. The concrete cover provides both chemical and physical barriers to corrosion.

The 'chemical barrier' is the high alkalinity of the concrete pore water solution, which ordinarily has a pH value of about 13. This enables the formation and maintenance of a permanent protective passivating film on the surface of the steel. The 'physical barrier' is the density and impermeability of the concrete cover, which limits the diffusion of oxygen toward the steel and thus may block the corrosion reaction, even if the passivating film has been disrupted.

Corrosion of the steel may begin if the chemical barrier ceases to be effective and the passivating film becomes unstable. This 'depassivation' can occur either in a limited area on the steel surface, usually around a crack, or else over a large region of the steel surface. The latter occurs usually when the concrete is porous and can be readily penetrated by carbon dioxide ($CO_2$). In the former case, corrosion is expected to be local, with the depassivated zone acting as the anode. When depassivation is more widespread, general corrosion is to be expected, with many active sites on the steel surface acting as anodes.

Once corrosion has begun, its rate will depend on several factors. The most important of these are: (i) the availability of oxygen at the cathodic areas; and (ii) the presence of water solutions in the concrete pores adjacent to the steel. This is necessary to provide a medium through which the ionic current from the cathode to the anode can flow.

In view of this two-step process of depassivation followed by active corrosion, Tuutti [4.1] proposed a model featuring the corresponding two stages, as shown schematically in Fig. 4.1. The concept is useful in understanding the influences of various parameters on the corrosion effects actually found in concrete [4.2, 4.3], and we shall discuss the initiation

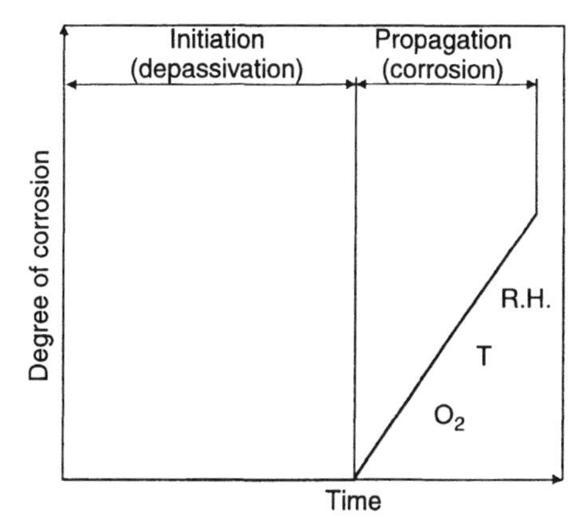


**Fig. 4.1.** Corrosion model for steel in concrete, consisting of two stages: initiation and propagation. **Initiation** is the result of depassivation by penetration of $Cl^-$ or by reduction in the pH due to carbonation ($CO_2$). **Propagation** is the stage at which the corrosion reaction is taking place and its rate can be controlled by availability of oxygen ($O_2$) or the electrical resistance of the concrete: the environmental conditions which control the rate at this stage are temperature (T) and relative humidity (RH). (After Ref. 4.1)

stage and the stage of active corrosion separately in this section. However, it should be kept in mind that the corrosion in the active corrosion stage does not ordinarily take place at a constant and uniform rate, as might be inferred from Fig. 4.1. Changes in environmental conditions, especially temperature and degree of wetness may accelerate or decelerate the actual rate of corrosion under practical conditions.

Another feature that should also be kept in mind is that corrosion damage in concrete is not limited to the reduction in steel cross-section. Such loss of section is the most important feature in most other corrosion areas, and it is thus most commonly used as a measure for rate of corrosion by corrosion engineers. In concrete however, most of the actual damage is in the form of cracking produced by the continued deposition of rust in the vicinity of the reinforcing steel within the concrete. The volume of the rust produced is substantially greater than the volume of steel that disappears during the corrosion reaction. In consequence, expansive stresses are set up in the concrete surrounding the steel. Cracking, and even spalling of the concrete cover may occur, and these responses may be far more damaging to the structure than the loss of section of the reinforcing steel. The process is somewhat self-accelerating, since once cracking begins the chemical and physical effectiveness of the cover is impaired, and higher corrosion rates may be expected.

In view of these considerations, it seems reasonable to propose a modified model for the corrosion processes as shown diagrammatically in Fig. 4.2. In this model the corrosion rate during the active corrosion or 'propagation' stage is shown to vary with time, and to be accelerated significantly when the concrete cover begins to crack.

## 4.2. THE CORROSION INITIATION OR DEPASSIVATION STAGE

The formation and stability of the passivating film on the steel surfaces is dependent on the pH level of the solution surrounding the steel. In the absence of interfering chloride or other ions, such a film has been shown to be produced and indefinitely maintained so as to effectively prevent corrosion when the pH is greater than about 11.5. Thus, the high pH concrete pore solutions is sufficient to maintain stable passivation films and corrosion should not occur.

However, in practice this favourable state of affairs may change and depassivation may occur under two specific sets of conditions: (i) reduction of the pH level due to reaction with atmospheric $CO_2$ (carbonation); and (ii) penetration of chloride ions into the concrete pore solution around the steel. It is known that, at sufficiently high concentrations, chloride ions can effectively destabilize the passivating film even when the normally high pH level associated with concrete water solutions remains unchanged. These two processes will be described in some detail in the following sections.

### 4.2.1. Depassivation by Carbonation

Carbon dioxide molecules that penetrate into the concrete can react with solid calcium hydroxide, with C-S-H gel, and with the alkali and calcium ions in the pore solution. The consequence of these reactions is a drastic

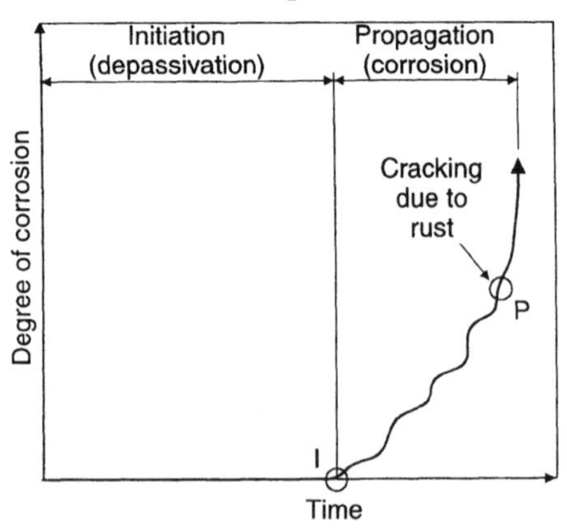


Fig. 4.2. Modified corrosion model for steel in concrete showing that the corrosion rates at the propagation stage are not constant and that they may be accelerated significantly once the concrete cover has cracked because of rust formation.

decrease in the alkalinity of the pore solution affected by the process, to levels of about pH 8. Usually a reaction front penetrates slowly into the concrete member. The time it takes the carbonation front to advance through the concrete cover and reach the surface of the reinforcing steel is a function of the depth of the cover, and of the rate of diffusion of atmospheric $CO_2$ into the particular concrete concerned.

The rate of this diffusion is controlled by the partial pressure of $CO_2$ in the surrounding atmosphere, and by the pore structure of the concrete. The partial pressure of $CO_2$ is usually considered to be constant, but it can be significantly higher than usual in certain industrial environments. The pore structure of the concrete is influenced only to a limited degree by the chemical composition of the Portland cement used, but it is very sensitive to the w/c ratio, the richness of the mix, and the effectiveness of the compaction and curing methods employed. It may also be influenced to a considerable degree by the incorporation of certain mineral admixtures, most notably silica fume, slag and fly-ash.

The rate of diffusion is also sensitive to the moisture content of the concrete. In saturated concrete the penetration of $CO_2$ can take place only by the diffusion of $CO_2$ that has been dissolved in the pore solution. Since, in common with that of most gases, the solubility of $CO_2$ in aqueous solution is small, comparatively little $CO_2$ can diffuse through water-saturated concrete. In practice, the rate of diffusion of $CO_2$ into air-dry concrete can be more than an order of magnitude greater than into the same concrete when saturated.

The extent of penetration of carbonation is usually expressed in length units, indicating the depth achieved by the carbonation front; rates of carbonation are therefore given in units of length per time. Carbonation rates through uncracked concrete can be as low as 1 mm/year if a suitably concrete mix design has been used and the concrete has been properly compacted and cured. Under such conditions, if the concrete cover is thick enough, the carbonation front would not reach a depth sufficient to depassivate the steel during the service life of the structure. This is the most important consideration in specifying a minimum depth of cover for a given structural member.

The diffusion of $CO_2$ into concrete will be discussed in more detail in section 4.3

### 4.2.2. Depassivation by the Effect of Chloride Ions

The influence of $Cl^-$ ions in depassivating steel even at high pH levels can be seen as a function of the net balance between two competing processes: stabilization (and repair) of the film by $OH^-$ ions, and disruption of the film by $Cl^-$ ions. It has been suggested that there is a threshold concentration of chloride ions that must be exceeded before depassivation can take place, and that this threshold concentration is a function of the pH, i.e. the $OH^-$ ion concentration. This suggestion has been generally accepted, but there is little agreement concerning the quantitative function relating the two. Hausman [4.4] suggested on the basis of measurements in $Ca(OH)_2$ solutions that the threshold $Cl^-$ ion concentration is about 0.6 times the $OH^-$ ion concentration.

The pH of calcium hydroxide solutions such as used by Hausman do not exceed 12.5; higher pH values can be found in concrete pore solutions due to the presence of dissolved potassium and sodium hydroxides. Gouda [4.5] made measurements in sodium hydroxide solutions in the high pH region, and on the basis of these results suggested another relation in which the threshold chloride ion concentration for pore solution of a given pH was expressed in a logarithmic form. Diamond [4.6] recalculated Gouda's data in terms of limiting $Cl^-/OH^-$ ratio, and found that at the higher pH levels which occur in practice a limiting ratio of approximately 0.3 rather than the 0.6 of Hausman might be more appropriate. The relationship between critical chloride ion concentration and pH implied by both the 0.6 and 0.3 criteria are given in Fig. 4.3.

Considerable uncertainty remains concerning the quantitative aspects of the relation between pH and $Cl^-$ ion threshold value. Both Hausman's and Gouda's data were derived from experiments in solution rather than in concrete *per se*, and other effects in concrete may influence the threshold value.

As a matter of practice, most specifications and guidelines do not address either the $Cl^-/OH^-$ ratio or the concentrations of these ions actu-

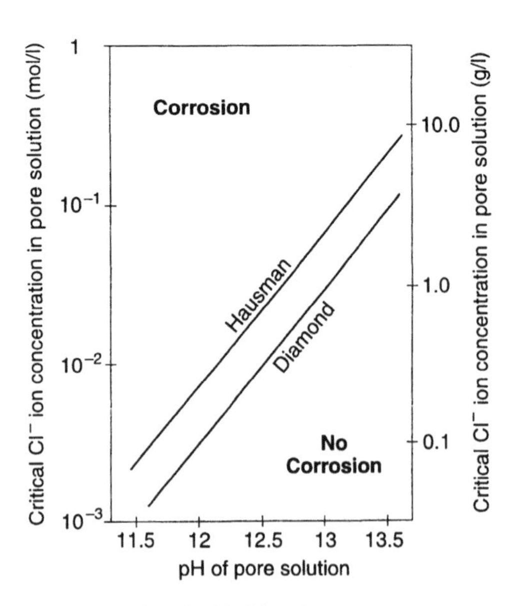


**Fig. 4.3.** Relations between threshold chloride ion concentration and the pH level of the pore solution as suggested by Hausman (Ref. 4.4) and Diamond (Ref. 4.6).

ally found in the concrete pore solution, but rather refer only to the total content of chloride in concrete. This is usually specified in terms of percentage by weight of the original cement used. Most specifications and recommendations that permit chloride at all recommend that the chloride content be smaller than about 0. 2% of the cement content of the concrete. In the range of 0.2 to 0.4% risk may be present, but not always. For example, the Building Research Establishment guidelines [4.7] suggest that for good-quality concrete, a $Cl^-$ ion content of less than 0.2% of the cement used should present only a low risk of steel corrosion; a high risk is considered to be present if the content is greater than 1%.

This range of values was confirmed by studies in which corrosion was characterized by electrochemical means. Relations between half cell potential measurements of embedded steel and the chloride content in the concrete at the steel level are presented in Fig. 4.4. The half cell potential can be used as a rough measure of the risk of corrosion (see Chapter 6); values less negative than –250 mV versus copper/copper sulphate (CSE) often represent conditions of low risk. Half-cell potentials more negative than –250 mV versus CSE were obtained when the chloride content exceeded 0.4% by weight of cement, indicating that this is a reasonable value to use for design as the threshold chloride concentration. Measurements of corrosion rates (Fig. 4.5) lead to similar conclusions: hardly any corrosion occurs below 0.4% chloride content; increase in corrosion rates starts at levels above 1% chlorides by weight of cement.

Sometimes the critical level of chloride is specified in terms of weight of chlorides per unit volume of concrete, in the range of 0.9 to 1.2 kg/m$^3$. This kind of specification is needed when assessing the chlorides in existing structures, where the total chloride content in units of weight per volume of concrete can be determined experimentally, but can not be calculated as % weight of cement, if the original composition of the

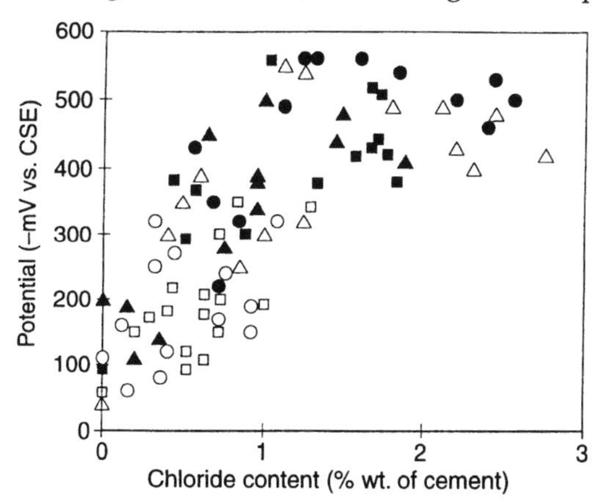


**Fig. 4.4.** Relationship between half-cell potentials and chloride contents. (After Ref. 4.8)

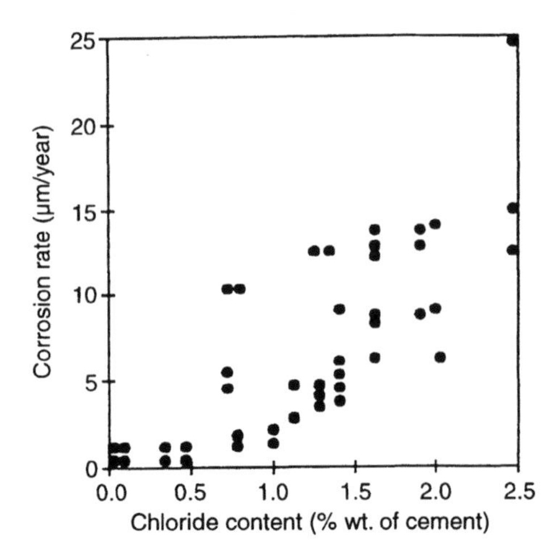


**Fig. 4.5.** Relationship between corrosion rates and chloride contents. (After Ref. 4.9)

concrete is not available. The range of 0.9 to 1.2 kg/m$^3$ corresponds to about 0.3% chlorides content by weight of cement, in concrete containing about 300 kg cement/m$^3$ of concrete.

The present specifications for chloride content recommended by the American Concrete Institute [4.10] are given in Table 4.1, which contains a different and more stringent limit for the chloride content in prestressed concrete than for conventionally reinforced concrete. This differentiation is common, since the consequences of corrosion in prestressed (or post-tensioned) steel are much more severe than in conventionally reinforced concrete. Indeed, ACI specifications and many others require that no calcium chloride-containing admixtures, regardless of the chloride level, be used in the production of prestressed concrete.

**Table 4.1.** Limits on Chloride Ions (% wt. of cement)[a]

| | | |
|---|---|---|
| 1. | Prestressed concrete | 0.06 |
| 2. | Conventionally reinforced concrete in a moist environment and exposed to external sources of chloride | 0.10 |
| 3. | Conventionally reinforced concrete in a moist environment but not exposed to external sources of chloride | 0.15 |
| 4. | Above-ground building construction where the construction will stay dry. Does not include locations where the concrete will be occasionally wetted, such as kitchens, parking garages and waterfront structures | No limit[b] |

[a]Suggestions by ACI Committee; Ref. 4.10.
[b]No limit for corrosion control. If calcium chloride is used as admixture, a limit of 2% is generally recommended for reasons other than corrosion.

The limits set on chloride content are ordinarily based on results of tests in which corrosion of reinforced concrete 'doped' with various levels of added chloride is determined. Results of tests of this kind are shown in Fig. 4.6 (unpublished results). In these tests there was hardly any sign of corrosion when the chloride content was less than about 0.3% by weight of the cement (equivalent to 0.5% of $CaCl_2$ by weight).

In general, the limits on chloride content as conventionally specified in terms of percentages of the cement used probably provide for satisfactory control of the problem. It should be taken into account that it is only chloride dissolved in the concrete pore solution that functions to depassivate the steel. The content of dissolved chloride in many circumstances can be significantly less than the total chloride content of the concrete, especially if cements rich in tricalcium aluminate and the ferrite phases have been used.

It is instructive to consider the various sources of chloride ordinarily found in concrete. Chloride can be introduced in the concrete mixing process, either as an admixture component (usually, but not always, as $CaCl_2$), or in chloride-contaminated aggregates or mix water. The chloride level of most Portland cements is low enough that chloride derived from Portland cement is usually not a factor of concern.

On the other hand, chloride ions can diffuse into the hardened concrete from external sources, such as sea water, salt spray, or de-icing salt placed on concrete pavements. The chloride concentrations that often develop in structural members above the water level in salt water exposures can actually be higher than that of the salt water itself, due to progressive concentration by repeated evaporation.

It should be appreciated that when soluble chloride salts are incorporated into a concrete mix, a substantial proportion of the chloride ions do not remain in the pore solution indefinitely, but are incorporated in solid cement hydration products. Data obtained by Diamond [4.6] indicate that for a typical cement at ordinary w/c ratios of the order of 0.5,

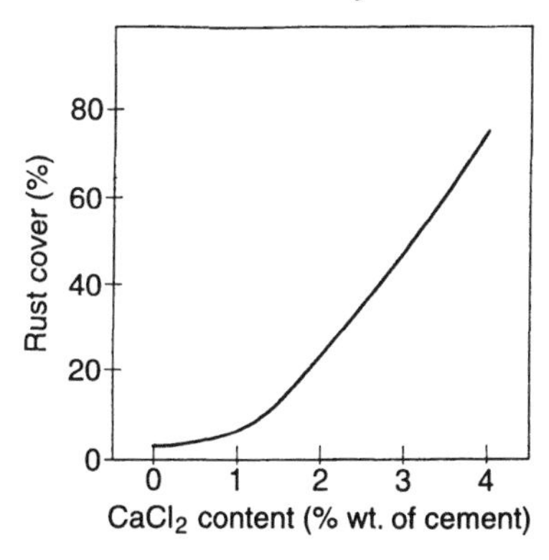


**Fig. 4.6.** Effect of $CaCl_2$ admixture content in the concrete on the extent of rusting of the reinforcement embedded in concrete piles, after 5 years. (After Ref. 4.11)

after a few weeks, the $Cl^-$ ion concentration in the pore water stabilizes at less than half of the starting value; significantly less if the starting concentration is low.

To some extent the proportion of chloride ions removed from the pore solution depends on the composition of the Portland cement. One of the hydration products that is produced in the presence of dissolved chloride ions is the relatively insoluble compound called Friedel's salt or tricalcium chloroaluminate, ($C_3A.CaCl_2.10H_2O$). The extent of formation of this compound depends on the $C_3A$ content in the cement. Thus, cements of high $C_3A$ contents inactivate a significant proportion of chloride that otherwise would remain in the pore solution and influence the onset of corrosion of steel.

Tests reported by Verbeck [4.12] have indeed shown that corrosion damage from chloride-induced corrosion is less for concretes in which the Portland cement used contained higher $C_3A$ contents, as shown in Fig. 4.7.

It has been shown [4.13, 4.14] that incorporation of silica fume into the concrete results in decreased removal of chloride ions from the pore solutions, presumably due to interference with the formation of the tricalcium chloroaluminate compound. Dissolved chloride ions may also be adsorbed on or incorporated in the structure of other hydration products produced in Portland cement concrete. As a practical matter, only the chloride ions remaining in solution are free to influence the passivating film directly, although it is possible that the less-available chloride in the system can act as a reservoir in situations where leaching of free chloride out of the concrete can occur.

When chloride ions are not derived from the concrete mix but diffuse into the concrete from the surrounding environment, the rate of diffusion is a function of the external chloride concentration, and of the

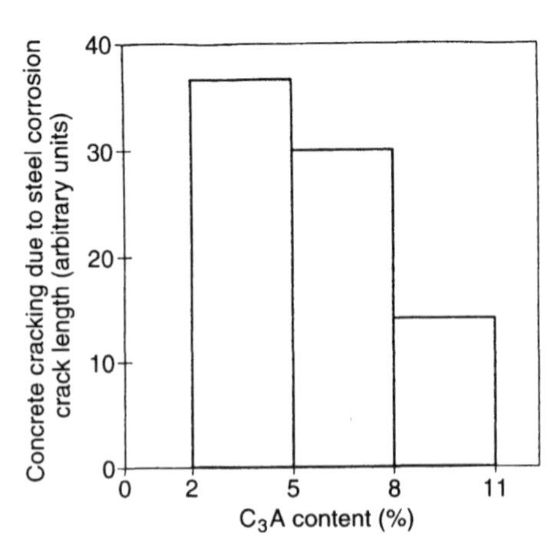


**Fig. 4.7.** Effect of $C_3A$ content in the portland cement on the corrosion damage (cracking of the concrete) due to corrosion of the steel in reinforced concrete exposed to sea water. (After Ref. 4.12)

density and pore structure of the concrete. In concretes of higher w/c ratio, i.e. more porous concretes, the rate of penetration of chloride ions is substantially greater than in dense concretes. The nature of the cement exerts some influence, and the incorporation of certain mineral admixtures such as silica fume and slag can lower the rate of penetration to very low values in suitably dense concretes. The diffusion rates of chloride ions in concretes of various kinds will be discussed in greater detail in section 4.3.

## 4.3. CONCRETE COVER: PROPERTIES AND THICKNESS

The time required to depassivate steel surfaces in reinforced concrete and thus start the corrosion reactions, and the rate at which these reactions will take place, both depend on the nature of the concrete cover provided.

Assuming that chlorides were not incorporated in the original concrete mix, the penetration either of chloride ions or of $CO_2$ molecules to the vicinity of the steel are controlled by the concrete cover. Specifically, they depend on the depth of cover, and the details of its properties, i.e. moisture content, extent of cracking, pore structure, the nature of the cement and any mineral admixtures used. Ideally, the time required to effect chloride or $CO_2$ penetration should be substantially greater than the life expectancy of the structure.

Once active corrosion has been started, its rate will also depend on the properties of the concrete cover. The rate may be limited by the availability of dissolved oxygen at the cathodic areas of the steel; in such cases the rate-controlling factor of the corrosion process will be the rate of diffusion of oxygen through the cover from the external atmosphere.

The concrete in the vicinity of the steel also must provide a path for the ionic current between the anodic and cathodic portions of the steel surface if corrosion is to be continued. The rate at which this current can transport ions may be the limiting factor in the rate at which corrosion occurs. This rate can be expressed in terms of the electrical resistivity of the concrete, which is a function of its physical and chemical characteristics.

In discussing the effects of the nature of the cover three different cases need to be considered:

(1) Uncracked concrete cover. Here, the diffusion rates and electrical resistance are controlled by the pore structure, the chemical characteristics of the concrete, and the degree of saturation of the pores.
(2) Cracked concrete cover. Here, the width and shape of the cracks become more important factors, and these control the rate of entry of chloride ions or the local rate of carbonation.

(3) Improperly compacted concrete cover. Here, the influence of cavities in the concrete and especially around the steel itself may dominate the entry of chloride and of $CO_2$.

Because of the practical importance of the subject, we will discuss each of these cases separately and in detail in the following sections.

### 4.3.1. Uncracked Concrete Cover

#### *4.3.1.1. Diffusion Processes and Diffusion Coefficients*

Uncracked concrete cover acts as a barrier to diffusion of $Cl^-$ ions and $CO_2$ gas (in controlling the corrosion initiation time) and of dissolved $O_2$ gas (in controlling the rate of corrosion once initiated). The diffusion processes of all of these species have some common characteristics, especially in that all are sensitive to the pore structure of the concrete. On the other hand, considerable differences do exist. Chloride must diffuse as ions in solution; the $CO_2$ diffuses almost entirely in the gas phase, and oxygen may diffuse either as dissolved oxygen molecules in solution or in the gas phase. Oxygen diffusing in the gas phase must then dissolve before it can take part in the cathodic reaction.

The penetration of chloride ions into the concrete can thus take place only through the aqueous solution within the concrete pores, or perhaps to some limited extent by surface diffusion along wet pore linings. The $CO_2$ and $O_2$ diffuse most easily through the gas phase in the pores of partly dry concrete, although dissolved molecules of oxygen do diffuse fairly extensively through water-filled spaces.

In order to place these concepts on a quantitative basis, one must consider the physical chemistry of the diffusion processes. While the following concepts are probably not overly familiar to most civil engineers and concrete technologists, they are not unduly complex, and should not be ignored.

Diffusion into concrete, or any substance, takes place only when the 'activity' of the dissolved ions or gas molecules in the external environment is greater than the activity of these species in the pores of the concrete. It is this difference in activity that provides the driving force for the diffusion. Activity in physical–chemical terms is a parameter related primarily to concentration. Thus, the driving force for the diffusion is the difference in concentration of the species outside of the concrete and that within the pores. Diffusion is usually described by Fick's law:

$$dc/dt = D\,(d^2c/d^2x) \qquad (4.1)$$

where $c$ is the concentration of the diffusing substance at a distance $x$ from the surface at time $t$, and $D$ is the diffusion coefficient of the process in units of $m^2/s$.

For the case in which the concentration of the diffusing substance on the exterior surface is effectively constant, as it is in these practical considerations, the concentration $C(x,t)$ within the concrete at a distance, $x$, after time $t$ will be given by:

$$C(x,t) = C_0 - \text{erf}\,[(x/2(Dt)^{1/2})] \tag{4.2}$$

where $C_0$ is the concentration at the boundary surface and erf is the error function. Using eqn. (4.2), the concentration profiles after various periods of diffusion can be calculated, and results of such calculations are shown schematically in Fig. 4.8.

For either chloride-induced depassivation or $CO_2$-induced depassivation, the depassivation will occur and active corrosion will commence when the concentration in the pores adjacent to the surface of the steel reaches the required critical value. Once corrosion has begun, its rate may be controlled by the flux of oxygen diffusing to the steel surface; if this occurs, the rate of consumption of oxygen in the cathodic reaction is the same as the rate of oxygen diffusing to these sites. As indicated in the equations above, this is a function of the diffusion coefficient $D$ for the process concerned.

The diffusion coefficient is not a constant, but depends on the temperature, the nature of the diffusing substance, and the nature of material through which diffusion is occurring. In concrete, diffusion takes place primarily through the pores of the hardened cement paste component, although it may also occur in the interfacial region between paste and aggregate. The diffusion coefficient should be sensitive, therefore, to the

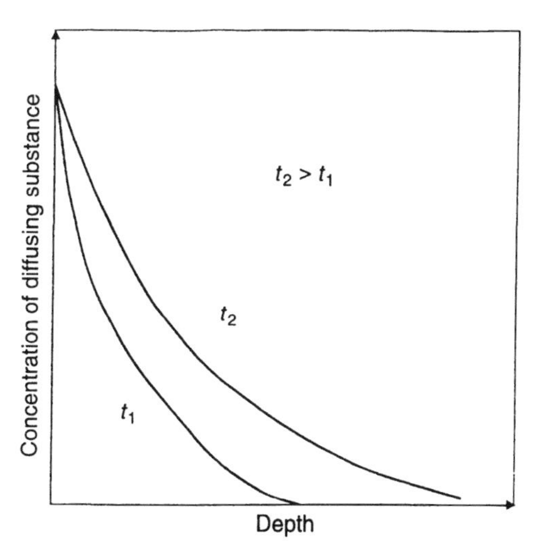


**Fig. 4.8.** Schematic description of the concentration profiles of a diffusing substance, at different times ($t_1$, $t_2$), assuming that its concentration on the surface of the concrete in contact with the atmosphere is constant.

pore structure of the concrete. For a dense concrete mixed at a low w/c ratio, the pores are primarily discontinuous and the diffusion path is probably very tortuous (Fig. 4.9(A)). Thus, the diffusion coefficients will have low values. For high w/c ratio concretes, the paste will be more porous and the pores larger and more extensively interconnected as shown schematically in Fig. 4.9(B); hence the diffusion coefficients will be accordingly larger.

The diffusion coefficient of the gases, $CO_2$ and $O_2$, will vary considerably depending on the moisture content of the concrete. A higher degree of water saturation in the pores will result in lower diffusion coefficients. The solubility of gas molecules in water is limited, and their mobility in water solution is much more restricted than their mobility in the gas phase. When a partially dry concrete is rewetted, water tends to fill small pores in between individual gel particles which can then act as bottlenecks with respect to gas diffusion, even if most of the capillary pore space remains empty. Thus, relatively small increases in moisture contents of concrete can result in large reductions of the rate of diffusion of $CO_2$ and of $O_2$ into the concrete.

For chloride ions and $CO_2$ molecules, the effective diffusion constant may be sensitive to interactions between the diffusing substances and the hydrated cement constituting the walls of the pores through which diffusion takes place. These interactions can be both chemical and physical. Diffusing chloride ions can be reacted chemically or can be adsorbed physically, both processes slowing down the diffusion process. Diffusing $CO_2$ molecules react chemically with calcium hydroxide and with C-S-H gel particles; the diffusion coefficient through concrete will therefore be significantly smaller than it would through an inert porous system of the same pore structure and degree of saturation.

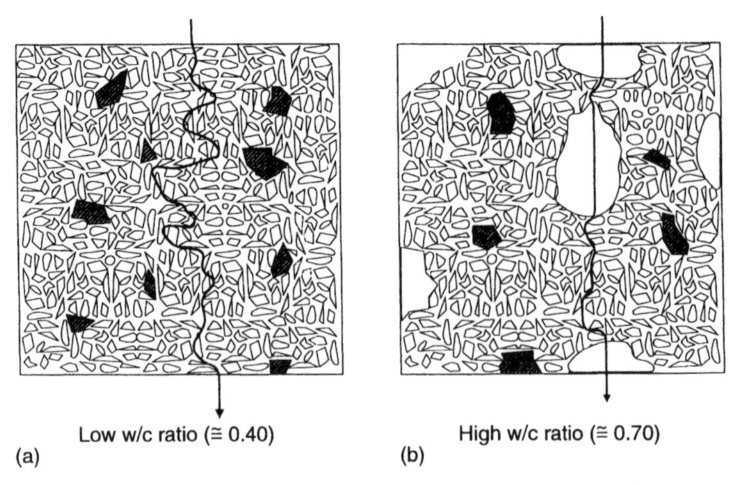


**Fig. 4.9.** Schematic description of the diffusion path in well-cured pastes of: (a) low w/c ratio (<0.40); (b) high w/c ratio (>0.60).

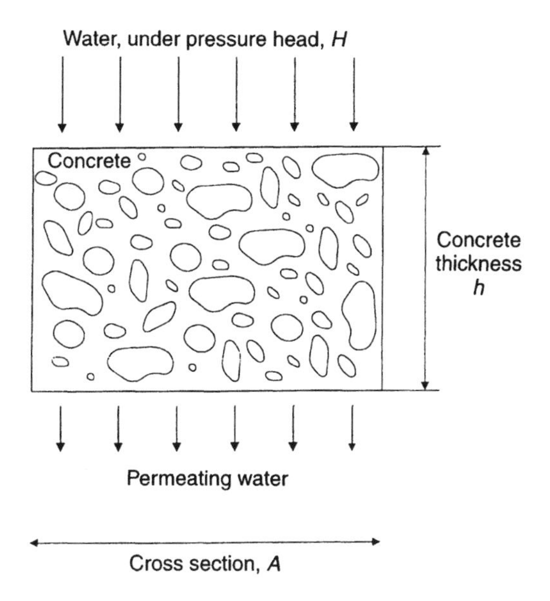


**Fig. 4.10.** Schematic description of water permeability test.

Sometimes the quality is assessed by determination of the permeability to the flow of water under pressure. Such permeability tests measure the steady-state flow rate of water under a constant pressure head through a representative concrete specimen, as indicated in Fig. 4.10. The coefficient of permeability, $K$, is calculated from D'Arcy's Law:

$$\mathrm{d}q/\mathrm{d}t = K\,A\,H/h \qquad (4.3)$$

where $\mathrm{d}q/\mathrm{d}t$ is the rate of water flow, $A$ is the cross-sectional area, $K$ is the coefficient of permeability in units of m/s, $H$ is the water pressure head in metres, and $h$ is the thickness of the concrete in metres.

The permeability coefficient is sensitive to the pore structure, as indicated in Fig. 4.11, derived from Powers *et al.* [4.15], which shows the relation between $K$ and w/c ratio. The high permeability coefficients observed for concrete at w/c ratios above 0.6 reflect the increasing volumes of interconnected capillary pores that remain unfilled by hydration products at progressively higher w/c ratios.

The dependency shown in Fig. 4.11 is qualitatively similar to those of the diffusion coefficients, providing some justification for the use of the permeability coefficient to characterize the quality of the concrete as a barrier to diffusion of ions and gases. However, it should be kept in mind that the diffusion coefficients for the different species of interest and the water permeability are only qualitatively related to each other; the driving force is different, and the water permeability measurement does not reflect the various physical and chemical interactions that the

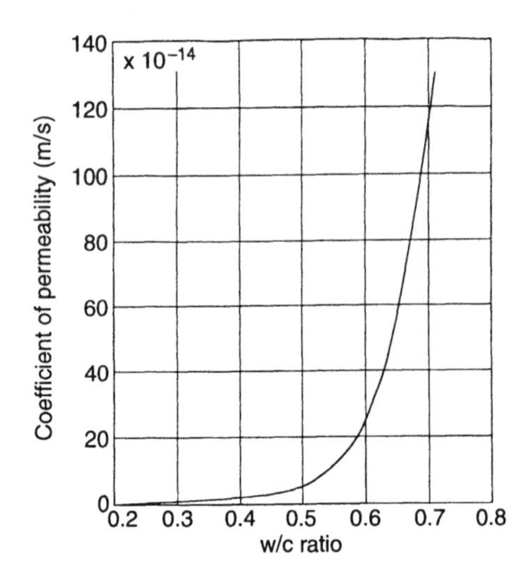


**Fig. 4.11.** Effect of w/c ratio of Portland cement paste on the coefficient of water permeability. (Adapted from Ref. 4.15)

diffusion species undergo with the pore walls and with the contents of pores.

Actually, the only situation in which the water permeability of the concrete is of direct concern in corrosion occurs when chloride-containing water (for example sea water) penetrates into dry concrete under pressure, as might occur in a concrete structure submerged after a coffer dam is removed.

Another mechanism by which chloride can penetrate into the concrete is by capillary absorption. This would occur in a structure which is partially submerged and intermittently exposed to chloride-containing water. The extent of chloride penetration by such effects would be particularly high in a high w/c ratio mix where capillary pores are interconnected (Fig. 4.12).

#### *4.3.1.2. Chloride Ion Diffusion through Uncracked Cover*

The diffusion coefficient of $Cl^-$ ions through concrete has been shown to be in the range of $0.1 \times 10^{-12}$ to about $25 \times 10^{-12}$ $m^2/s$. The effect of w/c ratio on this coefficient was determined by several investigators, and examples are given in Table 4.2.

The diffusion coefficient can also be quite sensitive to the type of cement used. It was demonstrated by Page *et al.* [4.16] and Berke and Hicks [4.17] that variations over an order of magnitude may exist in pastes and concretes prepared at the same w/c ratio, as shown in Table 4.3. These variations may reflect pore structure modifications as well as

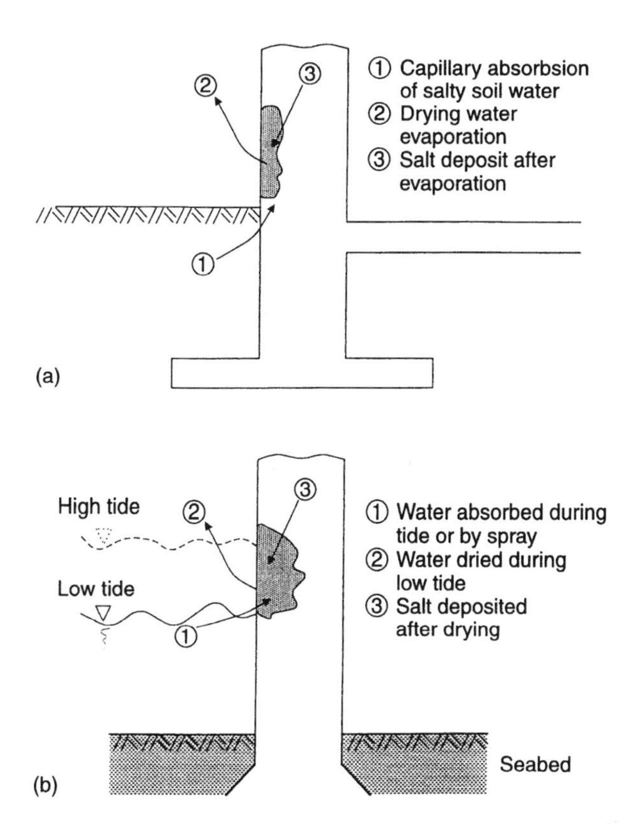


**Fig.4.12.** Penetration of salt containing water into a concrete structure by a capillary absorption process. (A) Penetration into a structure in contact with soil. (B) Penetration into a structure in the tidal zone.

specific pore wall chemical interactions that may tend to immobilize the diffusing chloride ions.

It is difficult to appreciate the full significance of the variations in diffusion coefficient without calculating concentration profiles at different ages. Such calculations have been made by Browne [4.3] for each of two concretes over 10- and 30-year diffusion periods. Browne assumed a

**Table 4.2.** Effective Diffusion of $Cl^-$ Ions in Cement Pastes and Concretes.[a]

| *w/c ratio* | *Diffusion coefficient (× $10^{-12}$ m²/s)* | | |
|---|---|---|---|
| | *Ref. 4.16*[b] | *Ref. 4.1*[c] | *Ref. 4.17*[c] |
| 0.4 | 2.6 | 0.8–5 | 2 |
| 0.5 | 4.5 | – | 11 |
| 0.6 | 12.4 | 8–12 | – |

[a] After Refs 4.1, 4.16 and 4.17.
[b] Pastes.
[c] Concretes.

**Table 4.3.** Effective Diffusion Coefficients of $Cl^-$ Ions at 25°C in Cement Pastes and Concretes of Different Composition at w/c of 0.5.[a]

| *Type of binder* | *Diffusion coefficient ($m^2/s$)* |
|---|---|
| Portland cement (OPC)[a] | $4.5\times10^{-12}$ |
| Sulphate-resistant Portland cement[b] | $10.0\times10^{-12}$ |
| OPC + 30% fly-ash[b] | $1.5\times10^{-12}$ |
| OPC + 65% blast furnace slag[b] | $0.4\times10^{-12}$ |
| OPC + 15% silica fume[c] | $0.7\times10^{-12}$ |

[a]After Refs 4.16 and 4.17.
[b]From Ref. 4.16.
[c]From Ref. 4.17.

chloride concentration at the external surface equivalent to 5% of $Cl^-$ ions by weight of cement, and diffusion coefficients of $0.1\times10^{-12}$ $m^2/s$ representing high quality concrete, and $5\times10^{-12}$ $m^2/s$ representing ordinary concrete. The resulting profiles are shown in Fig. 4.13.

From such curves, it is possible to estimate the time required for depassivation to occur at any given depth, if the threshold $Cl^-$ ion concentration for depassivation can be estimated. A reasonable threshold value estimate is 0.4% $Cl^-$ at the surface of the steel. Using this estimate, the data in Fig. 4.13 suggest that the cover thicknesses required for 30 years depassivation time are 50 and 100 mm for the high- and normal-

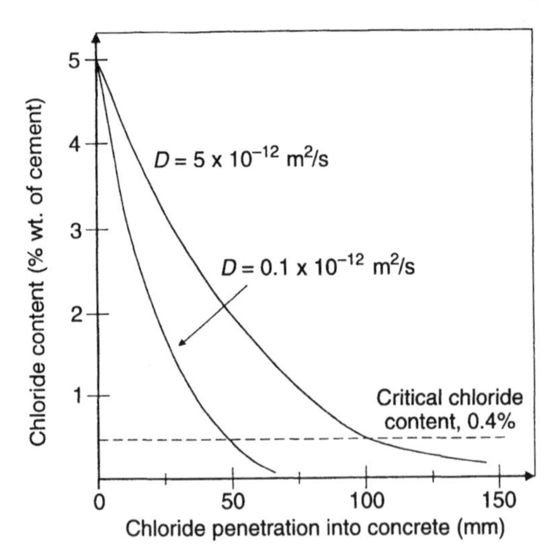


**Fig. 4.13.** Diffusion profiles of $Cl^-$ ions in normal and high quality concretes (diffusion coefficients of $5\times10^{-12}$ and $0.1-10^{-12}$ $m^2/sec$, respectively), after 30 years of diffusion, assuming a constant $Cl^-$ ion concentration of 5% by weight of cement, on the concrete surface. For a critical chloride level of 0.4%, concrete covers of 50 mm and 100 mm would be required in the high quality and normal concretes, respectively in order to prevent depassivation during 30 years of service. (After Ref. 4.3)

quality concretes, respectively. Note that this time period refers to the time necessary to depassivate the steel; corrosion will then begin, at a rate controlled at least in part by other factors, and the expected service life will include the time needed to corrode the steel to an extent that renders the structure unsuitable for further service.

The above example is representative of specific exposure conditions. A much more elaborate discussion of environmental effects and other engineering aspects of concrete will be given in Chapters 7 and 8.

While such calculations are useful in predicting at least a lower bound to the expected service life, one should keep in mind that they are valid only where the structure in question is continuously submerged in chloride-containing water. Often the real exposure conditions are much more complex. For example, structures along the sea shore may not be in direct contact with sea water but may be exposed intermittently to sea water splash, the water droplets containing the chloride (Fig. 4.12(B)). Subsequent drying of the splashed concrete causes a progressive build-up of the chloride on the surface of the concrete, sometimes to very high levels. As a result, when the concrete is rewetted, the rate of diffusion of the chloride ions toward the steel may be faster than expected. However, diffusion will only take place appreciably while the concrete is wet. The resulting balance of opposing factors on the actual penetration over a period of years is difficult to predict.

#### *4.3.1.3. Diffusion of $CO_2$ and $O_2$ through Uncracked Concrete*

The diffusion coefficient of gases through uncracked concrete is rather larger than that of chloride ions, values ordinarily being cited in the range of $0.1\times10^{-8}$ m$^2$/s to about $10\times10^{-8}$ m$^2$/s for $O_2$ diffusion. The specific value is naturally dependent on the pore structure and on the extent of pore water saturation. Tuutti [4.18] suggested that the coefficient of diffusion of $CO_2$ and $O_2$ are similar.

The effect of w/c ratio is illustrated in Fig. 4.14 (after Tuutti [4.18]) for diffusion in partly dry concrete at 50% relative humidity (RH). The effective diffusion coefficient through 0.8 w/c concrete is almost an order of magnitude higher than through 0.4 w/c ratio concrete under these conditions.

The dependence on the extent of pore water saturation is illustrated in Fig. 4.15, where diffusion coefficients are plotted as functions of RH for concretes of w/c ratio 0.7 and 0.4. Decreasing the ambient relative humidity with which the concrete is in equilibrium from 100% RH to 50% RH increased the diffusion coefficient through the 0.4 w/c concrete more than an order of magnitude, and through the 0.7 w/c concrete more than two orders of magnitude. It is not known to what extent these increases represent the results of changes in the interconnections of the pores on drying that may remain permanent on rewetting.

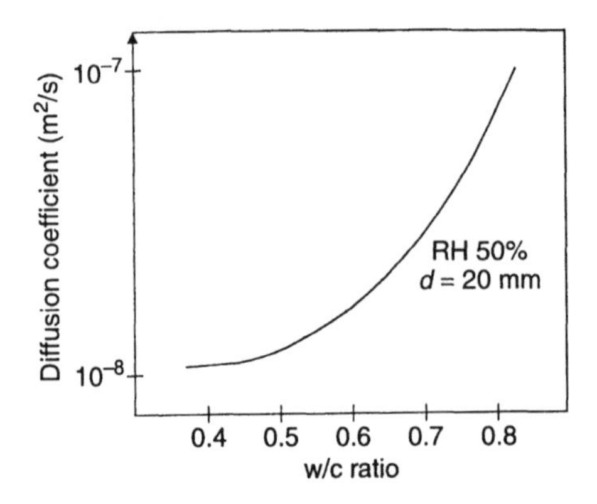


**Fig. 4.14.** Effect of w/c ratio on the diffusion coefficient of oxygen in paste at 50% relative humidity. (Adapted from Refs 4.1 and 4.18)

The practical consideration involved in oxygen diffusion through concrete cover are somewhat different from those concerning $CO_2$. The oxygen transport processes are of consequence not in terms of the time necessary to induce depassivation, as in the former case, but rather in terms of possibly limiting the ongoing rate of corrosion once corrosion has begun. What is of interest here is the ongoing rate of flow of oxygen to the cathodic areas of the steel surface. This is dependent not only on the diffusion coefficient, but on the depth of the cover provided.

Results of calculated flow rates for different cover thickness and for RH values of 60% and 80% RH in a 0.5 w/c ratio concrete are shown in Fig. 4.16, after Browne [4.3]. It can be seen from Fig. 4.16 that if oxygen flow is the limiting factor, at 60% RH increasing the depth of cover from 20 mm to 50 mm will decrease the rate of corrosion by almost an order of magnitude.

The sensitivity of such rates to RH is also apparent in Fig. 4.16. For a concrete cover of 30 mm, the data indicate that a decrease in RH from 80%

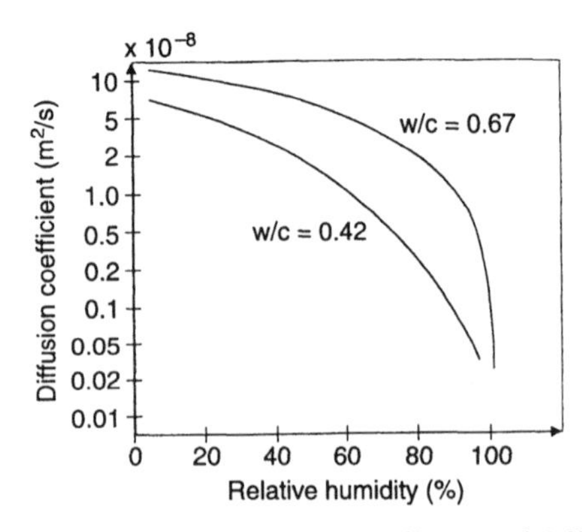


**Fig. 4.15.** Effect of relative humidity on the coefficient of diffusion of oxygen in 0.42 and 0.67 w/c ratio pastes. (Adapted from Refs 4.1 and 4.18)

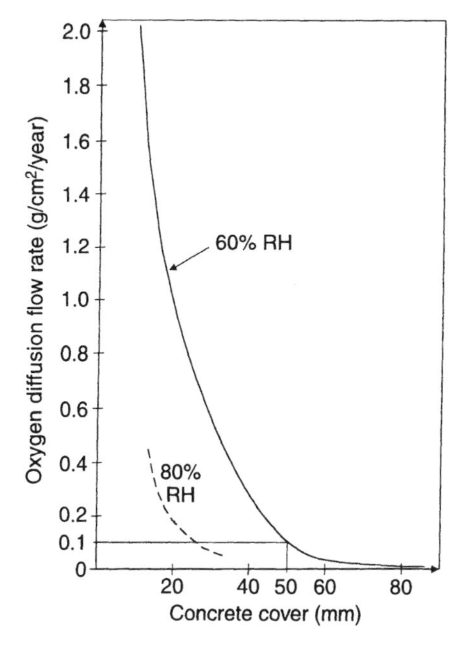


**Fig. 4.16.** Oxygen diffusion through non-saturated 0.5 w/c ratio concrete of different thickness and moisture content. (After Ref. 4.3)

to 60% will result in an order of magnitude increase in the expected rate of corrosion, providing that the steel is in a moist surrounding (RH > 70%).

### *4.3.1.4. Carbonation through Uncracked Concrete*

Carbon dioxide diffusing through uncracked concrete reacts with solid constituents and with alkalis in the pore solution to reduce the pH level from a highly alkaline value of approximately 13 to nearly neutral values. Since here it is the reduction in pH level that results in depassivation, one does not ordinarily attempt to test for $CO_2$ content as a function of depth; rather it is the depth of the zone of highly altered concrete that is usually measured. This is commonly done by spraying a phenolphthalein solution uniformly onto a freshly prepared concrete section normal to the surface of the concrete. The solution, which is colourless in high pH environments, changes colour to a deep purplish red when in contact with pore solution of pH greater than about 9. Thus, the current position of the incoming carbonation front is estimated visually from the position of the colour change.

The diffusion of gaseous $CO_2$ through the concrete cover is generally controlled by the same factors that influence the diffusion of oxygen. However, the partial pressure of $CO_2$ gas in the atmosphere is much lower than that of $O_2$ gas; in consequence the amount that can be dissolved is limited. Gaseous $CO_2$ will hardly penetrate into saturated concrete. As the moisture content of the concrete is reduced on drying and the interconnected pores become progressively emptied of fluid, the diffusion rate increases. The rate is also conditioned by the pore structure, as controlled primarily by the w/c ratio of the concrete.

There are, however, some special considerations. The effects of $CO_2$ penetration in inducing depassivation depend on the reduction in pH, which in turn depends on chemical reaction. The chemical reaction rate is limited under dry conditions; indeed, $CO_2$ will hardly react with completely dry concrete. In consequence, the most severe RH conditions from the point of view of depassivation by carbonation are those corresponding to intermediate moisture contents in the concrete pores, i.e. moisture contents low enough to permit ready penetration by the $CO_2$ gas, but high enough so that there is sufficient water for $CO_2$ chemical reaction.

In testing for the rate of penetration of the carbonated zone into concrete, it is usual to express the results in terms of the depth of the carbonated zone plotted against the logarithm of time. Such plots are usually linear, and the curves can be described empirically by eqn. (4.4):

$$x = \mathrm{k}\, t^{1/n} \tag{4.4}$$

where $x$ is the depth of the carbonated zone, $t$ is the time of exposure (ordinarily expressed in months or years), and k is a constant which depends on the effective diffusivity of $CO_2$ in the concrete, the ambient $CO_2$ concentration, and the reactivity of the cement paste with $CO_2$ under the moisture conditions prevailing. The inverse exponent, $n$, is ordinarily about 2; values in the range of 1.5 to 2.5 have been reported, however.

Typical plots showing the effects of w/c ratio and of various exposure conditions are shown in Fig. 4.17, after Tuutti [4.1]. Concretes sheltered

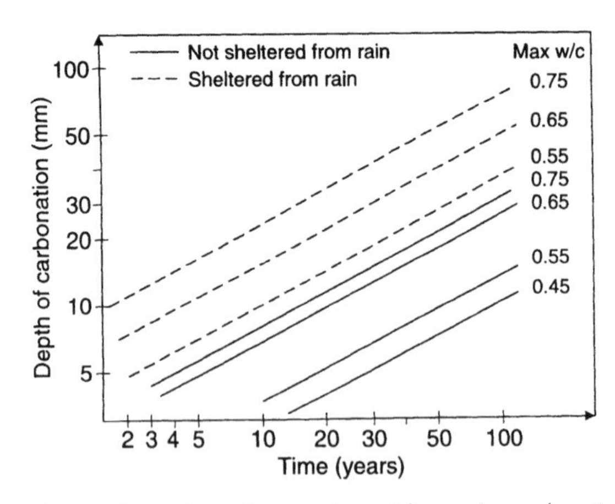


**Fig. 4.17.** Depth of carbonation of concretes with varying w/c ratio exposed to two different environmental conditions: sheltered and not sheltered from rain. (After Ref. 4.1)

from rain typically have an internal RH less than 90%, and are therefore more accessible to penetration of $CO_2$ than concrete that is exposed to the weather. The depth of carbonation is ordinarily greater for concretes of higher w/c ratio and consequent greater porosity (Fig. 4.17).

Such relations form the basis for the various current specifications which require concrete cover of thickness in the range of 25 to 75 mm, depending on the aggressiveness of the exposure conditions. In many specifications there are additional requirements regarding the quality of the concrete, set forth in terms of minimum strength, maximum w/c ratio, and minimum cement content. The logic behind such requirements can be readily understood from the data presented in Fig. 4.18. As an illustration, the expected depth of carbonation after 50 years in sheltered conditions is indicated as being reduced from 50 mm in 0.75 w/c ratio concrete to about 22 mm in 0.55 w/c ratio concrete.

Increasing the quality of concrete and the depth of cover will also have qualitatively similar effects in increasing the time for penetration of $Cl^-$ ions to the level of the steel, and in decreasing the rate of diffusion of oxygen to the steel. These technological aspects of the concrete cover will be further discussed in Chapters 7 and 8.

#### 4.3.1.5. *Electrical Resistivity of Uncracked Concrete*

After depassivation has occurred and the actual corrosion reactions have begun, an ionic current will be induced in the pores of the concrete cover

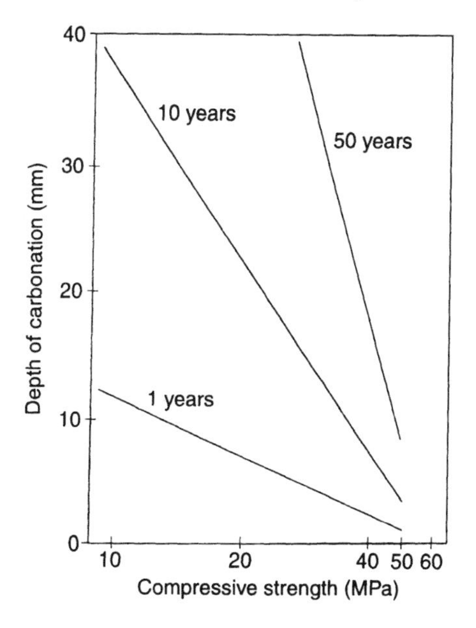


**Fig. 4.18.** Effect of concrete compressive strength on the depth of carbonation at different ages. (After Ref. 4.19)

surrounding the steel. As indicated in Fig. 2.1, the current is carried by $OH^-$ ions diffusing from the cathode toward the anode, and by $Fe^{2+}$ ions diffusing from the anode toward the cathode. The resistance to the flow of such current, i.e. the electrical resistivity of the concrete, is primarily influenced by three factors: the concrete pore structure, the moisture content or degree of saturation of the concrete, and the composition of the pore solution.

Concretes of low w/c ratio and consequent low pore volumes naturally have fewer channels available for ion transport and will have higher electrical resistivity than concretes of higher w/c ratio. Besides the effect of total pore volume, low w/c concretes tend to have smaller pores and pores which are more easily isolated by growth of hydration products at critical interconnections; thus, the electrical resistivity is further increased over what might be expected from the effect of reduced porosity alone.

The pore solution provides the primary medium for the ion transport. This solution usually has a high enough concentration of alkali and hydroxide ions to be a satisfactory electrolyte. However, if the concrete dries out or self-desiccates and the water content and degree of saturation of the pores are reduced, the pore spaces are partly emptied and the difficulty of ion transport is consequently multiplied. This is true even though the actual concentration of electrolyte in the remaining solution may be increased. Under these reduced ion mobility conditions the electrical resistance is increased substantially.

If the electrical resistance of the concrete is high, the extent of current flow per unit time is low, directly limiting the rate of corrosion [4.20, 4.21].

Specific indications of the effects of w/c ratio and of internal RH on the electrical resistivity of concretes are provided in Fig. 4.19, after Tuutti [4.1]. The data given indicate the great sensitivity of resistivity to the moisture level of the concrete. An increase in resistivity of about an order of magnitude occurs when the RH changes from 100% to 50%. The w/c ratio has an effect of similar magnitude, especially at intermediate water saturation levels.

Figure 4.20, after Browne [4.3], shows the specific effect of increasing the $Cl^-$ ion concentration in the pore solution. This added electrolyte reduces the resistivity somewhat, but the effect is rather small in view of the existing electrolyte already present. For a given concrete an increase in resistivity is indicative of a decrease in RH and/or decrease in concrete porosity. Both of these factors are beneficial in reducing corrosion.

#### *4.3.1.6. Corrosion Rates in Uncracked Concrete Cover*

The rate of corrosion occurring for steel in the active corrosion stage (i.e. after depassivation), assuming uncracked concrete cover, will depend primarily on two factors: (i) the availability of oxygen at the cathodic areas of the steel; and (ii) the moisture content. These two are not totally

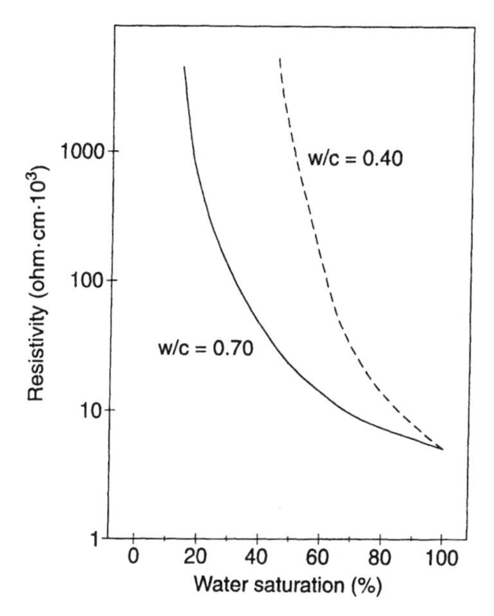


**Fig. 4.19.** Effect of water saturation and w/c ratio on the electrical resistance of concrete. (After Ref. 4.1)

independent factors, as described above. Thus, if the oxygen supply is limiting, increasing the degree of saturation by wetting or by exposing the concrete to higher relative humidity reduces the diffusion of oxygen,

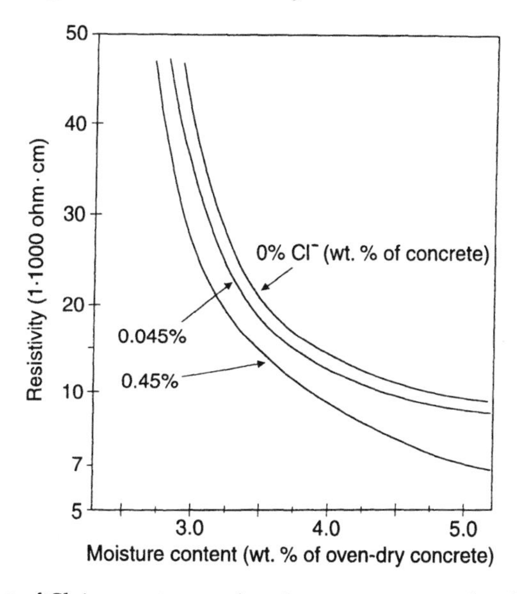


**Fig. 4.20.** Effect of $Cl^-$ ion content and moisture content on the electrical resistance of concrete. (After Ref. 4.3)

and thus leads to a lower rate of corrosion. However, increasing the degree of saturation decreases the electrical resistivity of the concrete. If the rate of corrosion is limited by the flow of the ionic current, increasing saturation can result in increased corrosion rate. There appears to be a critical internal RH, below which corrosion rates are negligible. On the other hand, in moisture-saturated concrete, corrosion rates can be limited by oxygen diffusion. This is shown schematically in Fig. 4.21.

The ambient temperature also exerts a significant effect on the rate of ongoing corrosion. In common with most chemical reactions, increasing the temperature increases the rates of reaction. The diffusion coefficients are also higher at higher temperatures, thus increasing the rates at which the reactants can be supplied to the reaction sites. As a rule of thumb, a 10°C increase in temperature doubles the corrosion rates. Influences of temperature will be further discussed in Chapters 7 and 8.

The considerations cited above indicate that, once active corrosion is underway, the corrosion rate of steel in concrete is likely not to be constant, but to vary with changes in moisture condition of the concrete and with temperature. Studies indicate that even at constant temperature, specimens subject to wetting and drying cycles corrode at rates that vary by more than an order of magnitude in different parts of each wetting and drying cycle. Such variations provide a severe limitation on our ability to predict practical rates of corrosion of steel in concrete subjected to the unpredictable variations in temperature and humidity characteristic of natural exposure, once depassivation occurs. This is illustrated in Fig. 4.2.

The difficulty is further compounded because of the slow adjustment of the internal RH in the concrete pores to changing external conditions. For example, when a concrete structure saturated by rain is exposed to drying conditions, the rate of evaporation of water from within the saturated pores is slow, and very long times elapse before the moisture content in

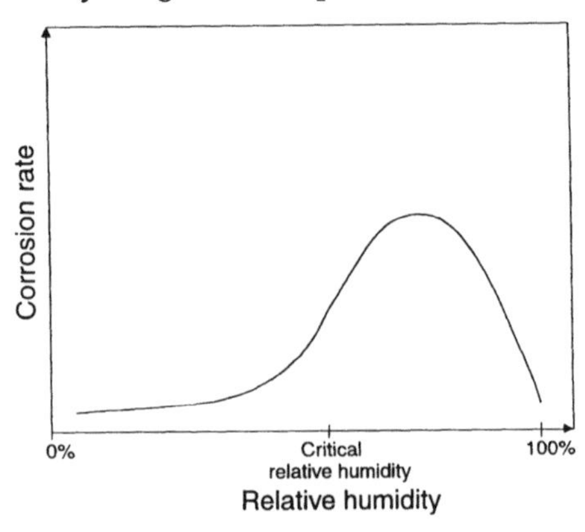


**Fig. 4.21.** Schematic description of the effect of relative humidity on the corrosion rate. Critical relative humidity could be in the range of 70–85% RH.

the pores is reduced to the level that represents equilibrium with the external RH. During this long adjustment period a moisture gradient must exist within the concrete, with near-surface pores being drier than those deeper in the concrete. This topic will be discussed further in Chapter 7.

The factor of self-desiccation of the concrete must also be considered. A number of studies, particularly that of Nielsen [4.22] suggest that concrete made at w/c ratios as high as 0.5 and not subject to external drying still develop and support internal RH values of the order of 80% except for the outer few millimetres of section depth.

### 4.3.2. Cracked Concrete Cover

Load-induced cracking is an inherent characteristic of reinforced concrete structures, since the maximum tensile strain that the concrete can support without cracking must usually be exceeded to use the steel reinforcement efficiently. In practice, only in special circumstances do design specifications call for the elimination of cracks. This can be achieved in prestressed or post-tensioned concretes.

Additional cracks often develop during service as the result of drying shrinkage and thermal effects. Further in-service cracking may be produced in concrete exposed to aggressive environments, especially where multiple freezing and thawing, sulphate attack, or alkali–aggregate reactions are significant factors. Consequently, concrete cover must often be expected to be cracked to a greater or lesser extent, and in considering the practical effect of the cover on corrosion, one must also consider the effects of cracks in the cover.

To make the problem still more complicated, the corrosion process itself often results in cracking of the cover, sometimes severe cracking. The extent and effects of corrosion-induced cracks will be considered separately in the subsequent section on corrosion damage.

Cracks in concrete cover produce at least three specific effects: (i) they tend to facilitate the onset of corrosion by providing easy access for the penetration of dissolved $Cl^-$ ions and of $CO_2$ so as to induce depassivation; (ii) they accelerate the rate of corrosion once begun, by reducing the barrier to diffusion of oxygen, at least near the cracks themselves; and (iii) they produce substantial non-uniformity in the physical and chemical environment around the steel, thus providing conditions which can lead to the development of 'concentration cell' galvanic corrosion.

It might be expected that the adverse effects of cracks would be proportional to their widths, since wider cracks should provide easier access for aggressive substances. This concept is the basis for the limitations on crack width specified in different codes of practice. The limiting crack width usually specified is of the order of 0.1 to 0.4 mm, depending on the nature of the aggressive environment expected. Typical values specified by ACI [4.23] are given in Table 4.4.

**Table 4.4.** Maximum Crack Width in Reinforced Concrete.[a]

| *Exposure conditions* | *Maximum crack width (mm)* |
|---|---|
| Dry air or protective membrane | 0.40 |
| High humidity, moist air, soil | 0.30 |
| Sea water and sea water spray, wetting and drying | 0.15 |
| Water-retaining structures, excluding non-pressure pipes | 0.10 |

[a]Recommendations of ACI Committee 224; Ref. 4.23.

The specifications included in the different codes provide equations to predict the crack width and crack spacing resulting from applied loads. These take into account the geometry of the concrete cross-section (depth of cross-section, depth of cover, percentage of reinforcement and bar diameter) and the properties of both the concrete and the steel. The cracks ordinarily considered are those developed perpendicularly to the main reinforcement.

The maximum permissible crack widths set by the various codes are based on experience in different countries, on the results of laboratory experiments, and to a much smaller extent, on theoretical analysis of the effect of crack width. The fact is that in many cases the correlation between crack width and extent of steel corrosion is not very marked. For example, Martin and Schiessel [4.24] showed that such a correlation was found for up to 4 years of exposure, but after 10 years of exposure no significant relation was observed.

A comparison of the behaviour of concretes meeting ACI minimum guidelines for crack width and concrete quality with the performance of reinforced concrete having high w/c ratio, lower cover and larger cracks was reported by Berke *et al.* [4.25]. It clearly showed the advantage of better quality concrete and smaller allowable crack width.

After critical review, Beeby [4.2] presented an analysis indicating that the main effect of the crack is to provide easy access of $CO_2$ through the concrete cover to the concrete surrounding the steel in the vicinity of the crack. This results in the development of a non-uniform carbonation front in the vicinity of the crack, as shown schematically in Fig. 4.22.

Schiessel [4.26] suggested the following equation relating the depth of carbonation near a crack, time, and crack width:

$$y = \mathrm{k}\,(wt)^{1/2} \tag{4.5}$$

where $y$ is the depth of carbonation in mm, $w$ is the crack width in the same units, $t$ is the time, and k is a constant depending on the concrete and on the exposure conditions.

The existence of cracks also increases the penetration of $Cl^-$ ions into concretes that are exposed to chloride-containing water or spray. In such cases, depassivation of the steel near the crack will take place before it occurs in uncracked areas. Corrosion then begins, with the depassivated

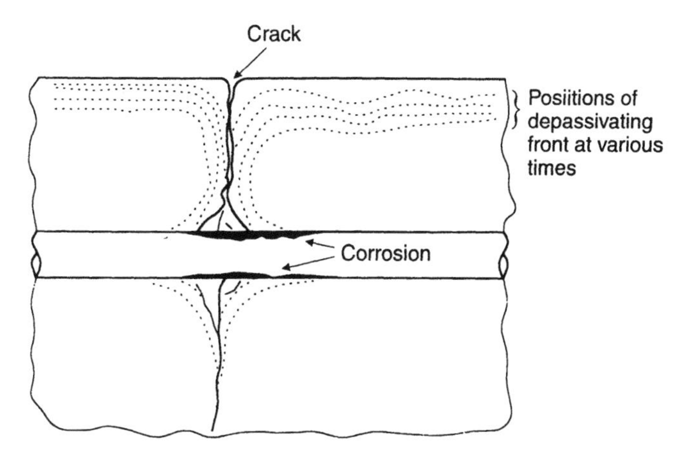


**Fig. 4.22.** Schematic description of the advance of depassivating front in a region of a crack. (After Ref. 4.2)

steel in the cracked zone being the anode and the remaining steel being cathodic, as indicated in Fig. 4.22. Subsequently the rate of corrosion will be controlled either by the diffusion of oxygen through the uncracked cover to the cathodic areas, or by the electrical resistance of the concrete around the steel. Thus, the subsequent corrosion rates are limited by the same factors as they are for uncracked concrete, and should not be substantially different from what they would be were the concrete not cracked.

The overall effects of cracks of differing widths are described schematically in Fig. 4.23, for two identical reinforced concretes that differ only in crack width. The time required to initiate corrosion, i.e. the depassivation time, is substantially shorter for concrete with the wider cracks (concrete 'A'), but the rates of corrosion after depassivation are not influenced significantly by crack width, and are shown to be similar in the figure. Thus at time $t_2$ there is already corrosion in the concrete with wide cracks, but none in the concrete with narrow cracks. At time $t_3$ there is corrosion in both concretes, and the extent of corrosion is not very different. Thus, observations made at time $t_2$ would show a strong correlation between crack width and extent of corrosion, but observations made at time $t_3$ would suggest that the influence of crack width is smaller.

Some additional geometrical aspects of cracks of varying widths must also be considered. One important factor not yet mentioned is the obvious fact that the width of the crack is ordinarily measured at the outer surface of the concrete; this is not necessarily its width on the inner surface in contact with the reinforcing steel. Cracks are not generally parallel-sided, but normally taper irregularly in width with distance inwards toward the steel. Another factor is the fact that cracking usually produces some local slip between the reinforcement and the concrete, resulting in additional separation between steel and the concrete in the vicinity of the crack. These characteristics are shown schematically in Fig. 4.24(A).

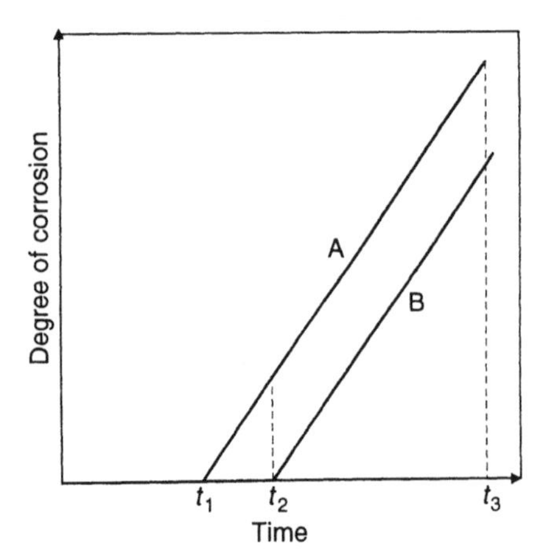


**Fig. 4.23.** Schematic description of the corrosion process in two identical reinforced concretes which differ only in their crack width. The width of the crack in concrete A is greater than that in B. $t_1$ is time to depassivation of concrete A; $t_2$ is time to depassivation of concrete B.

Some of the complications characteristic of the geometry of cracks in concrete cover are indicated by an example shown in Fig. 4.24(B), taken from Beeby [4.2], with the results of width measurements taken at various places indicated on the figure. It is apparent that the effective width of a crack in reinforced concrete cover may be very much narrower than the width exposed at the exterior surface.

Another factor not yet mentioned is the phenomenon of crack sealing. Sometimes corrosion products deposited in a crack can seal it effectively enough to prevent additional inward transport of substances which cause depassivation. If at the same time additional hydration products are produced and deposited around the steel in the vicinity of the sealed crack, the pH can increase sufficiently to re-passivate the steel. If this takes place active corrosion may be terminated. The sequence is indicated schematically in Fig. 4.25. This process was proposed by Schiessel [4.26] and discussed by Tuutti [4.1], who suggested that it might take place only in a carbonation-induced depassivation situation, and not where depassivation was induced by $Cl^-$ ions.

Sealing of cracks can also take place in submerged or wet structures where deposition of hydration products may occur sufficiently to seal the crack. These products may leach into the crack from surrounding areas, or they may be produced by additional hydration of incompletely hydrated cement near or within the crack zone itself.

#### 4.3.3. Defects In Concrete Cover other than Cracks

Defects in the concrete cover other than cracks which are of concern in corrosion may be classified as either defects within the concrete cover

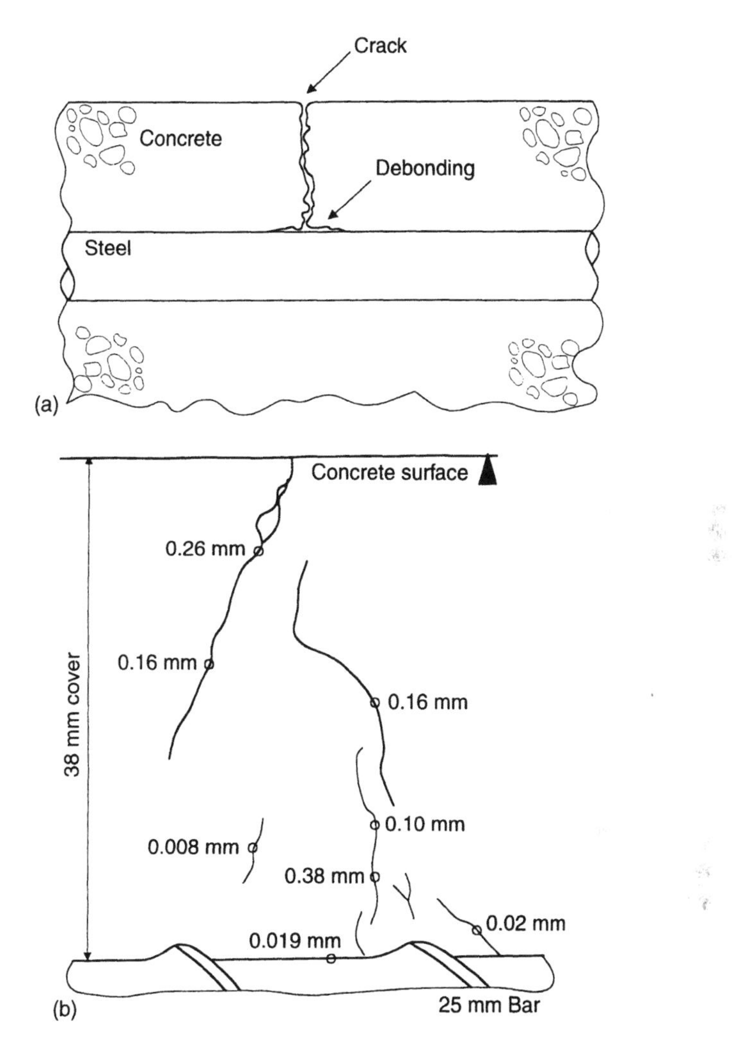


**Fig. 4.24.** (A) Schematic description of a crack in a reinforced concrete, showing that its width becomes smaller at the steel level, and the debonding at the intersection of the crack and the steel. (B) Actual crack in a reinforced concrete specimen. (After Ref. 4.2)

itself, or as separations between the steel and the concrete. Ideally, the mix design of the concrete should be such as to enable the formation of a dense layer of mortar in intimate contact with the reinforcing steel, as indicated in Fig. 4.26(A). However, skips or other defects in the cover may result from several causes. Inefficient compaction resulting from improper or inadequate vibration is not uncommon. Concrete mixes with too large a maximum aggregate size or an insufficient content of fine aggregate may result in honeycombing around the steel or other defects, as indicated in Fig. 4.26(B). Concrete that bleeds excessively produces

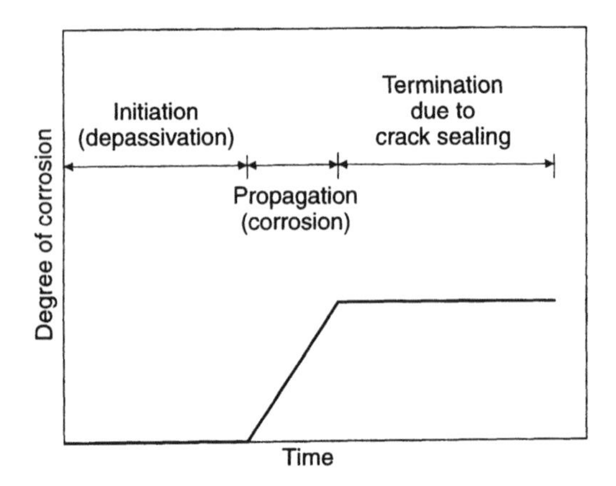


**Fig. 4.25.** Schematic description of the termination of corrosion process due to crack sealing.

water pockets that may become trapped under coarse aggregate pieces, and even under the reinforcing steel itself, as indicated in Fig. 4.26(C).

A common and often quite serious difficulty arises in post-tensioned concrete, where inefficient grouting of the ducts surrounding the reinforcing tendons leaves empty pockets. The steel in these empty areas is not covered with grouting mortar and is completely exposed to corrosion, as indicated in Figs 4.26(D,E).

The presence of skips or gaps in the cover substantially reduces its effectiveness as a barrier to the diffusion of $Cl^-$ ions and $CO_2$, and may thus result in early depassivation of the steel and the beginning of corrosion in the areas affected. Once corrosion has begun, its rate can be quite high, since the cover is also ineffective in slowing down the penetration of $O_2$. The availability of $O_2$ around the cathodic areas of the steel is often the rate-controlling factor in the corrosion process, especially in wet concrete.

Separations between the steel and the surrounding concrete may have even greater detrimental effects. Areas of steel that are not in direct contact with the concrete may not form passivating films, since the pH of the local solution may not be high enough. Thus, the corrosion reaction can start immediately; i.e. the time to initiation is reduced to practically zero. In such circumstances the steel out of contact with concrete will be anodic to the covered steel, and will corrode rapidly. The cell usually produced is a concentration-type corrosion cell of relatively high potential due to the great difference in the chemical environment of the steel out of contact with concrete and the steel that is effectively covered.

Effects of this kind are particularly severe in post-tensioned or prestressed concretes, where the stress maintained on the steel may make it sensitive to stress corrosion effects above and beyond the usual potential for corrosion.

## REFERENCES

4.1. Tuutti, K., *Corrosion of Steel in Concrete.* Swedish Cement and Concrete Research Institute, Stockholm, 1982.
4.2. Beeby, A.W., Concrete in the Oceans – Cracking and Corrosion. Tech. Rep. No. 2, CIRIA/EG, Cement and Concrete Association, UK, 1979.
4.3. Browne, R.D., Mechanisms of Corrosion of Steel in Concrete in Relation to Design, Inspection, and Repair of Offshore and Coastal Structures. In *Performance of Concrete in Marine Environment,* ed. V.M. Malhotra. ACI Publication SP-65, 1980, pp. 169–204.
4.4. Hausman, D.A., Steel Corrosion in Concrete. *Materials Protection,* **6**(11) (1967), 19–22.
4.5. Gouda, V.K., Corrosion and Corrosion Inhibition of Reinforcing Steel, I: Immersed in Alkaline Solutions. *British Corrosion J.,* **5**(9) (1970), 198–203.
4.6. Diamond, S., Chloride Concentrations in Concrete Pore Solutions Resulting From Calcium and Sodium Chloride Admixtures. *Cement, Concrete and Aggregates,* **8**(2) (1986), 97–102.
4.7. Anon., *The Durability of Steel in Concrete, Part I. Mechanism of Protection and Corrosion.* Building Research Establishment Digest No. 263, UK, 1982.
4.8. Bamforth, P.B. & Chapman-Andrews, J.F., Long Term Performance of RC Elements Under UK Coastal Exposure Conditions. In *Proc. Int. Conf. on Corrosion and Corrosion Protection of Steel in Concrete,* Vol. 1, University of Sheffield, July, 1994, pp. 139–56.
4.9. Podler, R. *et al,* Reinforcement Corrosion and Concrete Resistivity-State of the Art, Laboratory and field Studies. In *Proc. Int. Conf. on Corrosion and Corrosion Protection of Steel in Concrete,* Vol. 1, University of Sheffield, July, 1994, pp. 571–80.
4.10. ACI Committee 201, Guide to Durable Concrete, Chapter IV – Corrosion of Steel and Other Materials Embedded in Concrete, Manual of Concrete Practice, Part 1, The American Concrete Institute, Detroit, 1994.
4.11. Unpublished results, National Building Research Institute, Technion, Haifa, Israel.
4.12. Verbeck, G.J., Mechanisms of Corrosion of Steel in Concrete. In *Corrosion of Metals in Concrete,* ACI Publication SP-49, 1975, pp. 21–38.
4.13. Diamond, S., Effects of Microsilica (Silica Fume) on Pore Solution Chemistry of Cement Paste. *J. Am. Ceram. Soc.,* **66**(5) (1983), 682–4.
4.14. Page C.L. & Vennesland, O., Pore Solution Composition and Chloride Binding Capacity of Silica Fume Cement Pastes. *Materials and Structures,* **16**(91) (1983), 19–25.
4.15. Powers, T.C., Copeland, L.E., Hayes, J.C. & Mann, H.M., Permeability of Portland Cement Paste. *J. Am. Concrete Inst.,* **26**(3) (1954), 285–98.
4.16. Page, C.L., Short, N.R. & El-Tarra, A., Diffusion of Chloride Ions in Hardened Cement Paste. *Cement Concrete Res.,* **11**(3) (1981), 395–406.
4.17. Berke, N.S. & Hicks, M.C., Estimating the Life Cycle of Reinforced Concrete Decks and Marine Piles Using Laboratory Diffusion and Corrosion Data. In *Corrosion Forms and Control for Infrastructure,* ed. V. Chaker. ASTM STP 1137, American Society of Testing and Materials, Philadelphia, 1992, pp. 207–31.
4.18. Tuutti, K., Service Life of Structures With Regard to Corrosion of Embedded Steel. In *Performance of Concrete in Marine Environment,* ed. V. M. Malhotra, ACI Publication SP-65, 1980, pp. 223–36.
4.19. Smolczyk, H.G., Carbonation of Concrete – Written Discussion. In *Proc. Fifth Int. Symp. Chemistry of Cement, Part III,* Tokyo, 1968, pp. 369–80.
4.20. Browne R.D. & Baker, A.F., The Performance of Structural Concrete in a Marine Environment. In *Developments in Concrete Technology,* ed. F. D. Lyden. Applied Science Publishers, London, 1979, pp. 111–49.

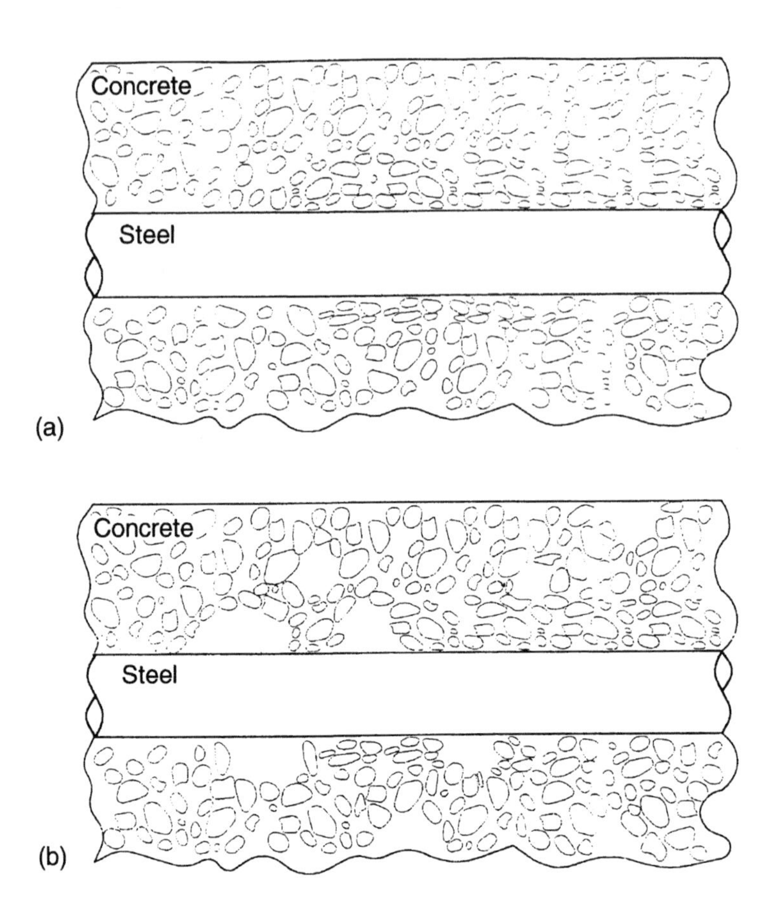
Concrete
Steel
(a)
Concrete
Steel
(b)

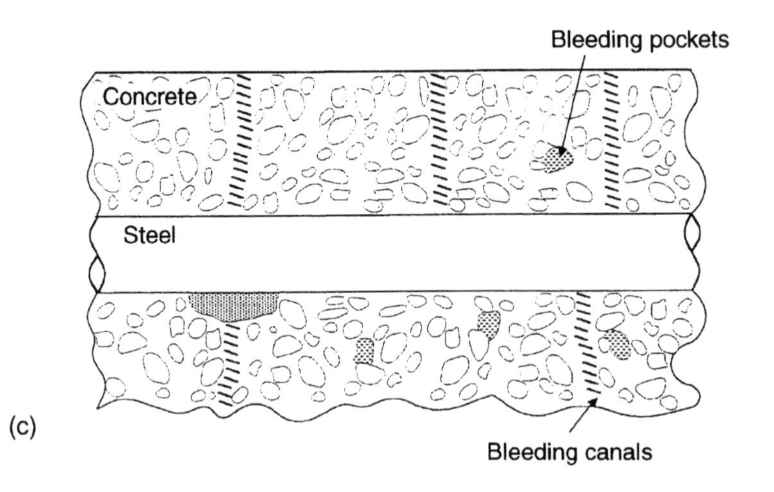
Bleeding pockets
Concrete
Steel
(c)
Bleeding canals

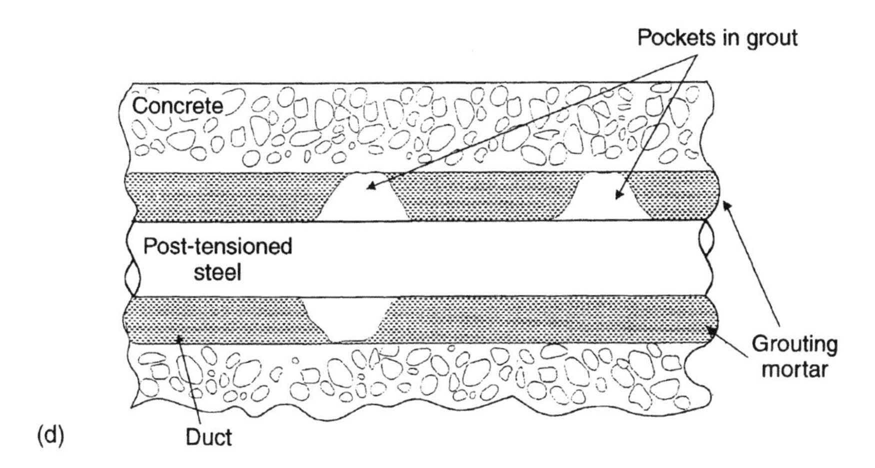


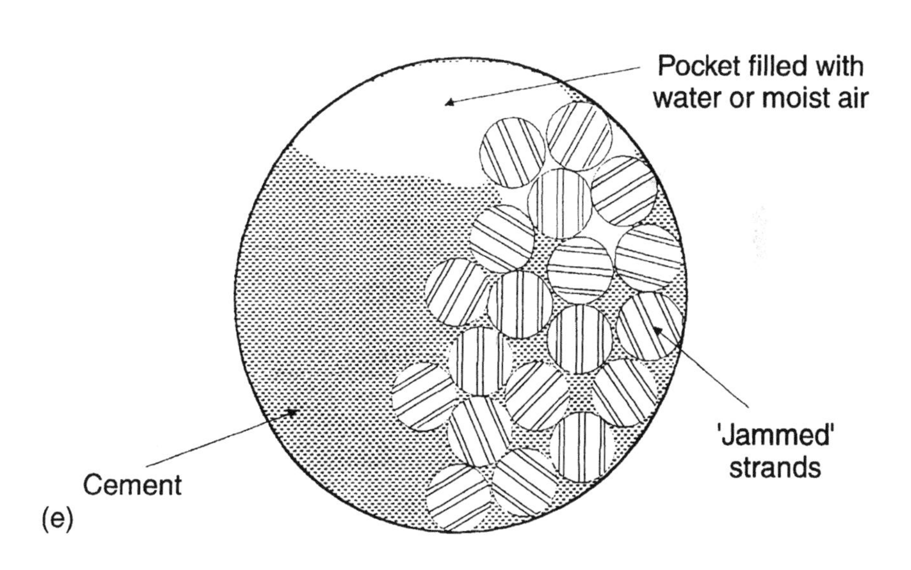


**Fig. 4.26.** Schematic description of the concrete cover. (A) Well-compacted concrete, showing mortar rich layer at the steel interface. (B) Poorly compacted concrete, showing void or honey combs in the concrete itself and at the steel interface. (C) Excessive bleeding in concrete cover showing trapped water underneath coarse aggregate and steel, as well as bleeding canals. (D) Pockets in grouting mortar in the duct around the post-tensioned steel. (E) Pocket in grout and ineffective filling of the spaces between the crowded tendons. (Adapted from Ref. 4.27)

4.21. Cavalier, P.G. & Vassie, P.R. *Proc. Inst. Civil Engineers*, **70** (1981), 462.
4.22. Nielsen, A., personal communication, 1985
4.23. ACI Committee 224, Control of Cracking in Concrete Structures. Manual of Concrete Practice, Part 3, 1994.
4.24. Martin, H. & Schiessel, P., The Influence of Cracks on the Corrosion of Steel in Concrete. In *Int. Symp. Durability of Concrete, 1969, RILEM, Prague*, 1969, pp. D205–18.
4.25. Berke, N.S., Dallaire, M.P., Hicks, M.C. & Hooper, R.J., Corrosion of Steel in Cracked Concrete. *Corrosion*, **49**(11) (1993), 934–44.
4.26. Schiessel, P., Admissible Crack Width in Reinforced Concrete Structures. Contribution No. II, Inter-association Colloquium on the Behavior of In-Service Concrete Structures, Preliminary Report. V.II, Liege, 1975, pp. 3–17.
4.27. Dykmans, M.J., Corrosion of Prestressing Steel in Concrete and How This Can Be Minimized Or Prevented? Paper presented at the Western Regional Conf., NACE, San Diego, 1976.

CHAPTER 5

# Corrosion Damage

In this section we discuss certain specific details of the damage in concrete structures that may result from different types of steel corrosion in concrete. This treatment is by necessity quite incomplete, and we hope that readers having responsibility for design and construction of reinforced concrete structures take every available opportunity to examine the results of corrosion in the field so as to improve their ability to recognize and evaluate the importance of such damage as it occurs in practice.

## 5.1. DAMAGE IN CONVENTIONALLY REINFORCED CONCRETE

The most direct damage resulting from steel corrosion in conventionally reinforced concrete is the reduction in steel diameter and cross-sectional area. The magnitude of the stresses carried by the remaining steel naturally increase. The risk to the safety of the structure will be a function of the residual difference between the increased stress level that results and the yield strength of the steel. This difference will be small, and a high safety risk will be incurred, only if the reduction in section is severe and the initial working stress of the steel was high.

Available data on the reduction in diameter due to corrosion in actual structures are rather scarce. However, its order of magnitude can be estimated on the basis of reports indicating that the maximum rates of steel corrosion in concrete are of the order of 50 μm/year [5.1, 5.2]. This is consistent with the results of corrosion testing [5.1, 5.3] which indicate that the corrosion current density is usually less than 10 $\mu A/cm^2$. A current density of 1 $\mu A/cm^2$ is equivalent to a rate of corrosion penetration into the steel of about 12 μm/year. From a practical point of view a corrosion rate of less than about 0.2 to 0.4 $\mu A/cm^2$ might be considered negligible; it is related to the current required to maintain a passive oxide (see section 6.2). However, if pitting corrosion occurs, as it may when the corrosion is initiated by chloride-induced depassivation, the local rate of penetration at the pit can be much greater.

The other type of damage done by corrosion in reinforced concrete is less direct, and more difficult to quantify. This is damage due to the formation of rust and its deposition within the concrete. Most of the rust is deposited in the vicinity of the corroding steel, but some of it can be leached out through the pores in the concrete cover, producing brown rust stains on the surface of the concrete. This process takes place more readily in wet concrete. The surface discoloration is one of the obvious signs that should be looked for when inspecting a structure for possible corrosion damage.

In general, the volume of rust produced in a corrosion reaction is at least twice the volume of the steel that is dissolved; thus, rust formation involves a substantial volume increase. Since most of the rust is confined within the concrete immediately surrounding the reinforcing steel, its formation generates expansive stresses within the concrete which can lead to cracking of the cover. As a result of the expansions produced, longitudinal cracks are often formed, and spalling and delamination can take place, as shown schematically in Fig. 5.1 (after Browne and Geoghegan [5.4]). It is apparent that the type of damage produced depends on the position of the reinforcing bar that is corroding. Cracks in the cover running parallel to the main reinforcement are often the first sign that the underlying steel is corroding.

Damage to the concrete cover from this source is extremely detrimental, since the effectiveness of the cover as a protective layer is markedly

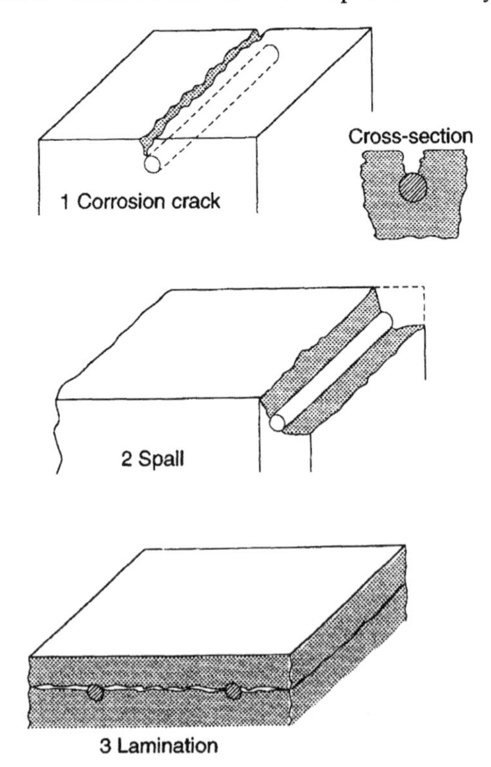


**Fig. 5.1.** Schematic presentation of the damage to the concrete cover due to the formation of rust. (After Ref. 5.4)

reduced. As a result, subsequent corrosion rates can be accelerated by a factor of 10 or more [5.2], and additional areas of reinforcing steel are opened up to easy access of $CO_2$ or $Cl^-$ ions, leading to subsequent depassivation and the start of new areas of corrosion.

The stage at which cracks in the cover are first observed depend not only on the amount of rust deposited, but also on the properties of the cover and the geometry of the situation. The strength of the cover, the cover thickness, and the diameter and location of the reinforcing bars are all important. Some of the relationships were summarized by Browne [5.5] and are presented schematically in Fig. 5.2.

There is always a possibility that in strong concrete a crack may form but be arrested within the concrete without necessarily propagating to the surface. A large depth of cover and a small diameter of reinforcing bar would favour this possibility; the ratio of the two values seems to be the governing parameter.

It has been suggested that the cracking of concrete cover ordinarily takes place when the thickness of the layer of rust deposited around the steel reaches 0.1 to 0.2 mm [5.5]. Such values are small compared with the usual reinforcing bar diameters. Thus, cracking of the cover would ordinarily be expected before the load-carrying capacity of the steel is drasti-

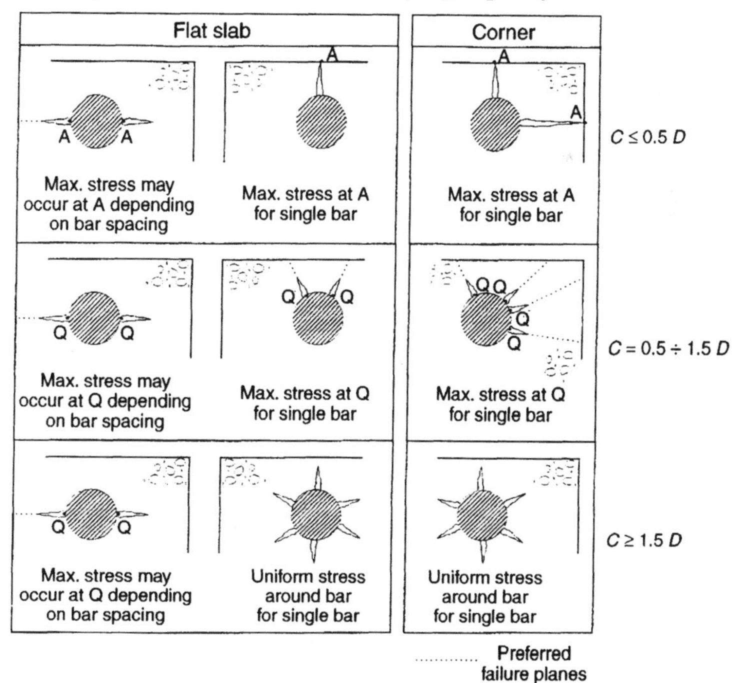


Fig. 5.2. Schematic description of the types of damage to the concrete cover due to the formation of rust, as a function of the position of the corroding steel bars and the ratio between depth of cover and steel diameter (C/D). (After Ref. 5.5)

cally impaired, and the first major corrosion damage observed is the cracking of the cover [5.2]. If necessary repair measures are not taken at this time, the corrosion rates may accelerate and lead to subsequent significant reductions in the steel cross-section which can produce considerable safety risks.

It should be noted that in some rare instances of low oxygen availability corrosion can occur, but will not result in expansive corrosion products. This is known as 'black corrosion' and is a serious condition since there is no visual warning as in the above case.

## 5.2. DAMAGE IN PRESTRESSED CONCRETE

The processes leading to corrosion damage in steel incorporated in conventionally reinforced concrete also affect steel in prestressed or post-tensioned concretes, but here the damaging effects may be more severe because of several factors specific to such concretes. Among these special factors are the following:

(1) The diameter of the steel tendons used in prestressed concrete is frequently much smaller than the diameter of reinforcing bars in conventionally reinforced concrete. For the same ongoing rate of corrosion, the relative thinning of the cross-section of the steel tendons in prestressed concrete will therefore be substantially greater.

(2) In post-tensioned concrete, incomplete or improper grouting can result in voids along the ducts, as indicated in Fig. 4.26(D). In such zones the steel tendons have practically no protection by concrete cover.

(3) Crowding of the tendons can prevent grout from penetrating between them, as indicated in Fig. 4.26(E). This results in lack of protection over the affected parts of the steel surfaces.

(4) A substantially increased potential for corrosion may occur where prestressing steel tendons make contact with a dissimilar metal, such as the duct wall or part of the anchorage. If the duct wall or the anchor are fabricated from galvanized steel, or indeed any metal different from that of the tendon, a galvanic corrosion cell may be induced. If the tendon steel is the more active of the two (although this seldom occurs) it will corrode; if less active, the other metal will corrode. The local rate of corrosion at such contacts may become especially severe if the grout around the contact is incomplete and pockets of air or water are present. There is always the danger that galvanic couple with the duct will result in hydrogen being produced on the tendon, leading to hydrogen embrittlement as discussed in the next section.

### 5.2.1. Ductile Versus Brittle Failure Modes

All of the corrosion processes considered so far can lead to ductile failure of the reinforcing steel if its cross-sectional area is reduced to such an extent that the stress level exceeds the yield strength of the steel. The affected tendons will start to deform plastically, and gradually fail. Subsequent examination of the steel after failure will show necking in the area of failure.

Corrosion-induced failure in prestressed concrete can, however, be of a different and sometimes much more catastrophic type, ordinarily associated with brittle materials [5.6]. Brittle failure of the steel may occur with only a little actual loss of metal. This type of failure can develop rapidly, in extreme cases in only a few weeks or months after tensioning. Such failures result from the simultaneous presence of tensile stresses and a corrosive medium. They can occur even in high-strength steels, including those customarily used for high-strength steel tendons in prestressed concrete. Such failures do take place more readily at higher stress levels; in consequence, adequate control of the actual stress levels attained in the tensioned tendons is important. Such stresses can be particularly high where the curvature of the wire profile changes, leading to the development of combined tensile and flexural stresses.

In point of fact, two different kinds of processes can lead to brittle failure resulting from corrosion damage. One is commonly called 'stress corrosion cracking', the other 'hydrogen embrittlement', or sometimes 'hydrogen cracking'. The modes of failure are quite similar, and it is difficult to distinguish between them by subsequent observation of the fractured steel surface. In either case a brittle failure without prior necking will be indicated.

Since a clear distinction between the two can often not be made, it has been suggested that some of the failures reported as having been caused by hydrogen embrittlement may have actually been induced by the stress corrosion mechanism [5.6, 5.7].

The environmental conditions associated with the two mechanisms are quite different from each other. Hydrogen embrittlement takes place in more acid environments, i.e. those in which the concentrations of hydrogen ions (protons) is high. Since the concrete pore water solution is always alkaline and has a correspondingly low concentration of protons, hydrogen embrittlement is likely only where the tendons are not protected by concrete or mortar, and especially if such unprotected areas are exposed to acidic environments such as may be produced in polluted industrial air. This can occur in practice in any of the following situations:

(1) In construction work where external prestressing steel that will subsequently be covered with shotcrete is not protected for long periods of atmospheric exposure before the actual gunning.

(2) Where tendons are ungrouted or improperly grouted so that ungrouted zone remain within the ducts, or where there is a substantial delay between stressing and grouting.
(3) Where improper design or construction practices result in unprotected tendons at joints and anchorages.
(4) Where cracks occur in the concrete cover around pockets in the grout, thus directly exposing the steel to the environment.
(5) Where improper handling and lack of protection of the steel occur during shipping and storage on site, resulting in prolonged exposure to acidic atmospheres and in accumulation of aggressive substances on the surfaces of the steel.

Hydrogen embrittlement can also occur if the strands are subjected to stray currents or when cathodic protection is improperly applied. In these conditions production of hydrogen at the surface can occur, even though the pH is high [5.8, 5.9].

### 5.2.2. Stress Corrosion Cracking

Failure in the stress corrosion cracking mode occurs as a consequence of the joint presence of relatively high stresses and of a corrosive environment, but the details of the mechanism responsible are not well understood. In simplified terms, this form of corrosion failure can be regarded as an extreme form of localized corrosion, associated with the formation of a micro-pit. The tip of the pit is subject to highly concentrated stress while it is at the same time undergoing dissolution as an active anode in a localized corrosion process, as indicated in Fig. 5.3.

Such conditions occur more readily in stressed metals that have previously been protected by passivating films. Once breakdown of the film occurs, a pit may form in the affected zone. The high stress concentration at the tip of the pit produces progressive plastic deformations that act to

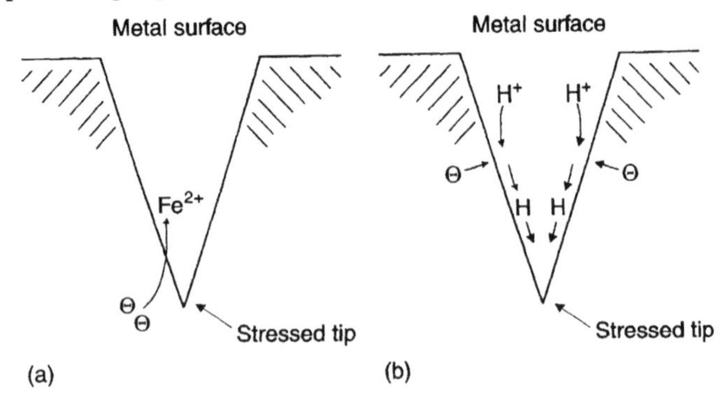


**Fig. 5.3.** Schematic description of the corrosion process taking place at the tip of a pit or a crack in a stressed metal leading to its embrittlement. (A) Stress corrosion cracking. (B) Hydrogen embrittlement.

prevent any repassivation. Thus, the tip continues to remain anodic, and its dissolution continues at a high rate. This combination of stress and rapid corrosion induces a crack that propagates rapidly, leading to macroscopically brittle fracture.

### 5.2.3. Hydrogen Embrittlement

Hydrogen embrittlement occurs when steel is under high stress and at the same time areas on its surface are involved in a type of cathodic reaction not previously discussed. This cathodic reaction evolves hydrogen atoms at the steel surface. Hydrogen atoms so produced can diffuse through to and dissolve in the most highly stressed zones of the steel, as shown in Fig. 5.3(B). The effect of the hydrogen accumulation in the stressed zone is to embrittle the steel.

The dissolution of hydrogen in the steel is possible because of the extremely small size of the hydrogen atom. It is facilitated in zones of the metal that are under high tensile stress, such as those present around any existing crack tips in the steel tendons.

The type of cathodic reaction mentioned above is not likely to occur in alkaline environments such as those of concrete pore solutions, but occurs characteristically in acidic environments. The reaction is given below:

$$\underset{\text{hydrogen ions}}{2H^+} + \underset{\text{electrons}}{2e^-} \rightarrow \underset{\text{adsorbed hydrogen atoms}}{2H^o} \quad (5.1)$$

In this process, two electrons produced during the anodic dissolution of iron (eqn. (2.la)) and transferred to the cathodic site, combine with two hydrogen ions at the surface of the metal in contact with the surrounding solution to produce two hydrogen atoms which are adsorbed on the surface. These atoms can either combine to form hydrogen gas molecules, $H_2$, or they can dissolve into the metal and diffuse away from the surface.

A number of specific details of the embrittlement mechanism are not fully understood [5.10], but it has been suggested that hydrogen atoms may collect at dislocations and interfere with their slip movements, thus interfering with the ability of the metal to deform plastically. When hydrogen migrates to the region of a crack tip, the fracture stress level at the tip may be reached before any yielding occurs due to this local inhibition of plastic yielding or slip, with the overall result being the onset of brittle fracture. High-strength steel alloys are particularly susceptible to this kind of cracking.

As indicated previously, hydrogen atoms formed on cathodic surfaces as the result of the processes indicated in eqn. (5.1) can either combine to form hydrogen gas or else diffuse into the metal as individual atoms of dissolved hydrogen. These two are competitive processes; which one is favoured depends considerably on the local environment. The presence of certain chemical species in the solution, particularly dissolved hydrogen sulphide, $H_2S$, or indeed of sulphide ions generally, will hinder the combination of hydrogen atoms to form gas molecules, and will favour the dissolution of hydrogen into the metal.

The failure of prestressed steel by hydrogen embrittlement is thus likely only when a combination of factors are present: when the tendons are imperfectly protected by concrete, when the surrounding environment is acidic, and especially when dissolved sulphides are present. Such exposures can occur for steel in concrete structures exposed to polluted industrial air, and also for concrete in contact with sewage or with certain industrial waste waters. Sources of sulphides can include surface condensates or deposits on steel during shipping or storage before incorporation in concrete, contamination of the concrete itself by extraneous components, and sometimes even by specific components deliberately incorporated into concrete mixes. Examples of the latter include some slags and sometimes high-alumina cement.

The importance of the acid environment has been stressed by several investigators [5.11–5.13], who have shown that even in the presence of $H_2S$, hydrogen embrittlement of high-strength steel occurs only if the pH of the surrounding solution is less than about 7 to 9. At first glance, this would rule out most concrete environments, but Isecke [5.14] has shown that, in practice, water trapped in ducts before grouting can have pH values as low as 7.

## REFERENCES


5.1. Tuutti, K., *Corrosion of Steel in Concrete*. Swedish Cement and Concrete Res. Inst., Stockholm, 1982.
5.2. Beeby, A.W., Concrete in the Oceans – Cracking and Corrosion. Tech. Rep. No. 2, CIRIA/EG, Cement and Concrete Association, UK, 1979.
5.3. Gonzales, J.A., Alonzo, C. & Andrade, C., Corrosion Rate of Reinforcements During Accelerated Carbonation of Mortar Made With Different Types of Cement. In *Corrosion of Reinforcement in Concrete Construction*, ed. A.P. Crane. Soc. Chem. Ind., UK, 1983, pp. 159–74.
5.4. Browne, R.D. & Geoghegan, M.P., The Corrosion of Concrete Marine Structures: The Present Situation. In *Proc. Symp. Corrosion of Steel Reinforcement in Concrete Construction*, Soc. Chem. Ind., UK, 1979, pp. 79–98.
5.5. Browne, R.D., Mechanisms of Corrosion of Steel in Concrete in Relation to Design, Inspection, and Repair of Offshore and Coastal Structures. In *Performance of Concrete in Marine Environments*, ed. V.M. Malhotra. ACI Publication SP-65, 1980, pp. 169–204.
5.6. Szilard, R., Corrosion and Corrosion Protection of Tendons in Prestressed Concrete Bridges. *J. Am. Concrete Inst.*, **66**(1) (1969), 42–59.
5.7. Heidersbach, R., personal communication, 1984.
5.8. Hart, W.H., A Critical Evaluation of Cathodic Protection for Prestressing Steel in Concrete. In *Corrosion of Reinforcement in Concrete*, eds C.L. Page, K.W.J.

Treadaway & P.B. Bamforth. Society of Chemical Industry/Elsevier Applied Science, UK, 1990, pp. 515–24.

5.9. Pangrazzi, R., Hart, W.H. & Kessler, R., Cathodic Polarization and Protection of Simulated Prestressed Concrete Pilings in Sea Water. Paper No. 92, Corrosion 92, The NACE Annual Conference and Corrosion Show, National Association of Corrosion Engineers, Houston, Texas, 1992.

5.10. Fontana, M.G., *Corrosion Engineering*. McGraw-Hill, 1986.

5.11. Hudgins, C.M., McGlasson, R.L., Mehdizadeh, P. & Rosborough, R., Hydrogen Sulfide Cracking of Carbon Alloy Steels. *Corrosion*, **22**(8) (1966), 238–51.

5.12. McCord, T.G., Bussert, B.W., Curran, R.M. & Gould, G.C., Stress Corrosion Cracking of Steam Turbine Materials. *Materials Performance*, **15**(2) (19976), 25–36.

5.13. Griess, J.C. & Haus, D.J., Corrosion of Steel Tendons Used In Prestressed Concrete Pressure Vessels. In *Corrosion of Reinforcing Steel in Concrete*, eds D.E. Tonini & J.M. Gaidis. ASTM STP 713, 1980, pp. 32–63.

5.14. Isecke, B., The Influence of Constructional and Manufacturing Conditions on the Corrosion Behaviour of Prestressed Wires Before Grout Injection. In *Corrosion of Reinforcement in Concrete Construction*, ed. A.P. Crane. Soc. Chem. Ind., UK, 1983, 379–92.

CHAPTER 6

# Corrosion Measurements

The measurement of the corrosion activity of steel actually occurring in a given concrete is an important tool in assessing the present and future performance of reinforced concrete structures exposed to corrosive environments. In addition, such measurements are extensively used in laboratories in the development of building codes and new corrosion protection systems.

Many measurement techniques can be used for studying metal corrosion in general. However, the opaque nature of the concrete cover, and the inability simply to remove the steel in concrete for separate study and examination limit the techniques that can be adopted for concrete corrosion studies. Furthermore, the fact that the resistivity of concrete is much higher than that of normal aqueous environments requires that various modifications to standard techniques are employed.

In this chapter the theory and practice behind the most commonly used measurement techniques for steel in concrete will be discussed. The interpretations that can be made will be reviewed, and potential pitfalls of overinterpreting corrosion data will be highlighted.

## 6.1. THEORY

Corrosion theory is complicated. Fortunately, it is not necessary to become an expert to understand the significance of the various techniques used to measure corrosion rates. Nevertheless, a basic understanding of some of the key principles is necessary, and these will be discussed, especially as to how they relate to steel in concrete, in addition to the treatment provided in Chapter 4.

As noted earlier, steel in non-carbonated concrete in the absence of chloride does not corrode. This is due to the highly alkaline environment that promotes the formation of a passive oxide film around the steel. Passivity occurs because a tightly adherent corrosion product is produced that has an extremely low dissolution rate and protects the steel below. Discussion of the passivation process in terms of the chemical reactions involved was presented in section 2.3. The passivation process can also be described in electrochemical terms. To do so we have to define a few simple terms used by corrosion engineers.

When a metal is placed into an aqueous solution it comes to an equilibrium or steady-state condition. This state is quantified by comparing the potential (voltage) difference between the metal and a reference electrode exposed to the same environment. The difference in potential is called 'corrosion potential'. The corrosion potential of steel in the environment of concern is measured, and related to the corrosion behaviour.

The reference electrode is fabricated from a material with behaviour that is virtually independent of the environment. Typical reference electrodes used for steel in concrete are copper/copper sulphate electrode (CSE), the saturated calomel electrode (SCE), and the silver/silver chloride (Ag/AgCl). When stating a potential it is necessary to state which reference electrode was used (e.g. –100 mV versus SCE).

The value of the corrosion potential is dependent on the activities of the cathode and anode. The chemistry of the anodic and cathodic reactions was presented in Chapter 2. The electrochemistry of these processes can be described in terms of curves of potential versus log electrical current density as shown schematically in Fig. 6.1. The current is the rate of the anodic (Fig. 6.1(A)) or cathodic (Fig. 6.1(B)) reactions. These curves are referred to as polarization curves, with the curve in Fig. 6.1(A) being the anodic polarization curve, and the curve in Fig. 6.1(B) being the cathodic polarization curve.

The anodic polarization curve for steel in an alkaline environment is presented in Fig. 6.1(A). At strongly negative potential, referred to as active potential, the corrosion rate increases rapidly with the development of more positive potential (points 1 to 2 in the curve in Fig. 6.1(A), and than suddenly drops by several orders of magnitude when passivation reactions occur to form ferric oxide. The potential at which this occurs is known as the primary passivation potential, $E_{pp}$ (points 2 to 3). Then, over a large range of potentials the steel is corroding at a negligible rate (points 3 to 4) and this is the passive region discussed in Chapter 2. At still higher potential, referred to as more noble, the breakdown of water to produce oxygen occurs and severe corrosion can once again occur and passivity is lost (point 5). The potential at which this occurs is the transpassive potential ($E_{tp}$). The cathodic behaviour is shown in Fig. 6.1(B). The rate of the cathodic reaction to produce $OH^-$ decreases with increase in the potential (from point 1 to 2 on the curve in Fig. 6.1(B).

As in other chemical processes the anodic reaction and the cathodic reaction have to be in balance. The electrons flow from the anodic to the cathodic sites, and there is a counterflow of negative ions through the solution from the cathodic to the anodic sites. These flows of electrons and ions are the anodic and cathodic currents, respectively. The corrosion potential, $E_{corr}$, is the potential at which the anodic and cathodic reaction rates are in balance, and as such the absolute values of the anodic and cathodic currents are equal.

This is shown in Fig. 6.1(C) in which the anodic and cathodic polarization curves have been superimposed and the rates have been shown as

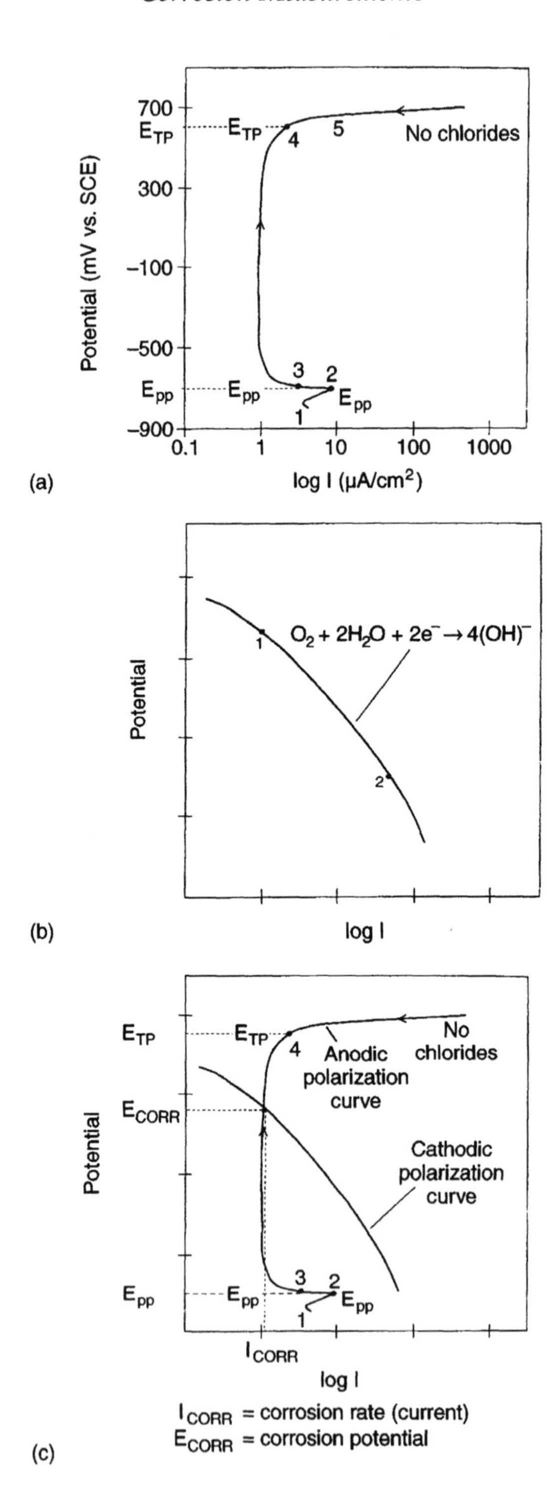

700
300
−100
−500
−900
$E_{TP}$
$E_{pp}$
Potential (mV vs. SCE)
No chlorides
1
2
3
4
5
0.1
1
10
100
1000
log I (μA/cm²)
(a)
Potential
$O_2 + 2H_2O + 2e^- \rightarrow 4(OH)^-$
log I
(b)
Potential
$E_{CORR}$
Anodic polarization curve
No chlorides
Cathodic polarization curve
$I_{CORR}$
log I
$I_{CORR}$ = corrosion rate (current)
$E_{CORR}$ = corrosion potential
(c)

currents. The current at the corrosion potential ($E_{corr}$) is defined as the corrosion current, $I_{corr}$. Since steel is usually passive in non-carbonated chloride-free concrete, the intersection of the two reactions is in the passive zone. Implications of changing oxygen content and/or the anodic behaviour due to chloride ingress will be discussed below.

Now that the passive behaviour of steel in concrete has been explained using electrochemical concepts, it will be shown how the same concepts can be used to describe the corrosion behaviour when either chlorides or carbonation is present.

The anodic behaviour of steel in concrete is drastically changed in the presence of chloride, as shown by the anodic polarization curve in Fig. 6.2(A). The presence of chlorides causes pitting to initiate. When pitting is present the corrosion current density in the pit is high and the corrosion current is only limited by the extent of cathode present. The anodic polarization curve is almost horizontal at the pitting protection potential, $E_p$ (point 5 in Fig. 6.2(A)). At potentials negative to $E_p$, pits are not stable and pitting stops. At potentials positive to $E_p$, the corrosion rate is orders of magnitude higher than the passive rate and severe local metal loss can occur.

Increasing the chloride content has the effect of facilitating the pitting process and results in a further lowering of the pitting protection potential as shown in Fig. 6.2(B). For the same extent of available cathode, the corrosion rate increases progressively with chloride content, as shown in Fig. 6.2(C). This is where the rule of thumb about more negative corrosion potentials being associated with higher corrosion rates came from.

The role of oxygen on the corrosion rate when pitting is present is illustrated in Fig. 6.3. In this figure we plot a polarization curve at a single level of chloride concentration as the anodic response. The effects on the cathodic polarization curve for four levels of dissolved oxygen are also presented on the same plot as curves I, II, III, IV (Fig. 6.3(A)). The extreme right hand curve (curve I) is for a high oxygen concentration. Under this condition the intersection of the cathodic curve with the anodic curve (point 1) (Fig. 6.3(B)) is at a relatively high $I_{corr}$ and a relatively low $E_{corr}$. At a somewhat lower oxygen content, the cathodic curve at low potentials declines nearly vertical (curve II). Nevertheless the intersection point does not change much (point 2), and the high $I_{corr}$ is maintained. On further reduction in dissolved oxygen content the cathodic curve shifts to curve III, and is displaced to the left of curve II over much of the poten-

**Fig. 6.1.** Polarization curves. (A) Anodic polarization curve of steel in alkaline solution (i.e. passivated steel). (B) Cathodic polarization curve representing the cathodic reaction $O_2 + 2H_2O + 4e^- \rightarrow 4(OH)^-$. (C) Schematic presentation of the intersection of the anodic and cathodic polarization curves. The two reactions are balanced at the intersection; the potential and current at the intersection point are defined as corrosion potential ($E_{corr}$) and corrosion current or rate of corrosion ($i_{corr}$)

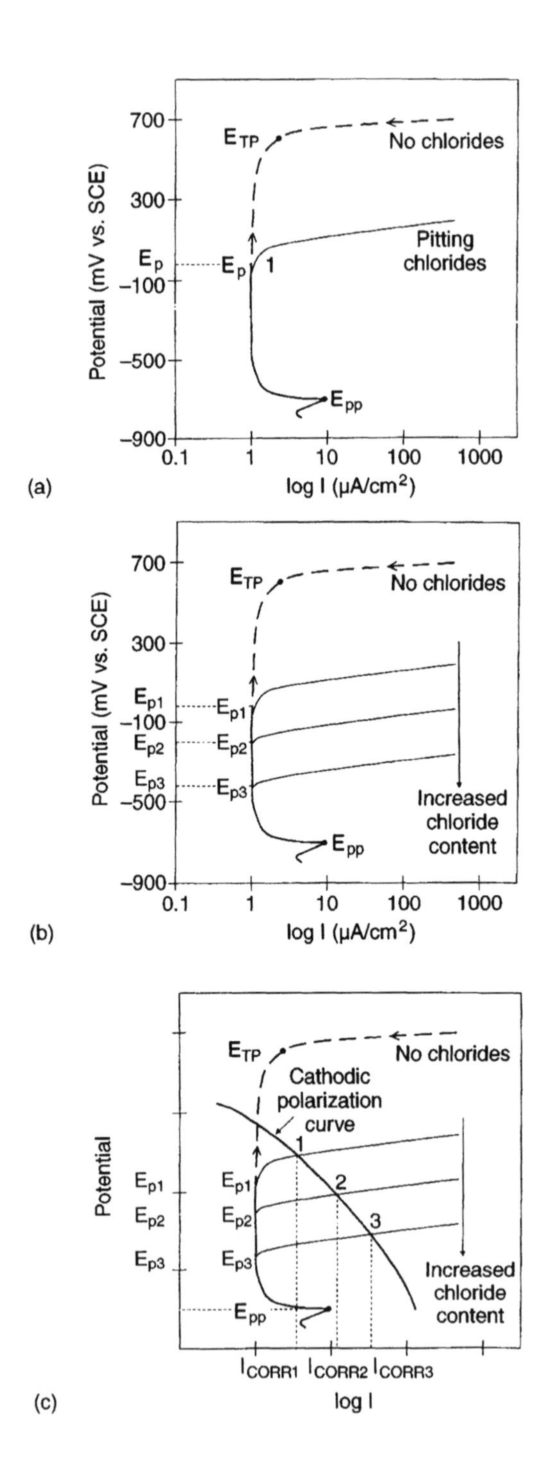
700
300
Ep
−100
−500
−900
Potential (mV vs. SCE)
ETP
No chlorides
Pitting
chlorides
Ep
1
Epp
0.1
1
10
100
1000
log I (μA/cm2)
(a)
700
300
Ep1
−100
Ep2
Ep3
−500
−900
Potential (mV vs. SCE)
ETP
No chlorides
Ep1
Ep2
Ep3
Epp
Increased
chloride
content
0.1
1
10
100
1000
log I (μA/cm2)
(b)
Potential
ETP
No chlorides
Cathodic
polarization
curve
1
2
3
Ep1
Ep2
Ep3
Ep1
Ep2
Ep3
Epp
Increased
chloride
content
ICORR1
ICORR2
ICORR3
log I
(c)

tial range. The intersection with the anodic curve is shifted drastically to lower $I_{corr}$ (point 3), without appreciably changing $E_{corr}$. When dissolved oxygen is reduced to extremely low levels (as may occur in submerged structures), the cathodic curve shifts even further to the left (curve IV). The intersection with the anodic curve (point 4) is now seen to be in the passive region and $I_{corr}$ shows negligible value.

When observations are made under conditions of varying oxygen concentrations, the drop in measured corrosion potential is associated with a reduction in $I_{corr}$. This is different from the trends in Fig. 6.2(C). In that case the oxygen level was kept constant but we compared different concentrations of chlorides. Under these circumstances the reduction in the corrosion potential was not associated with reduction in corrosion rate, but rather with continuous increase. This comparison emphasizes the need to assess the environmental conditions when drawing conclusions regarding corrosion rates which are based only on corrosion potential measurements.

Note that even if sufficient chloride is present for pitting, corrosion rates can be low if the oxygen content is low. This is commonly observed for piles far below the water line (sea water) marine environments. At higher oxygen levels, corrosion only occurs if the oxygen reduction reaction intersects the anodic reaction at above the protection potential (Fig. 6.2).

When carbonation is present, the steel is depassivated and as a result the anodic polarization curve is different than the passivated one, as shown in Fig. 6.4(A). This results in a significantly higher corrosion rate which is dependent upon the oxygen content, as shown in Fig. 6.4(B).

## 6.2. LABORATORY CORROSION MEASUREMENTS

Corrosion measurements are used to determine the corrosion rate or corrosion mechanisms. They can be as simple as making visual observations, or involve sophisticated electrochemical testing techniques and equipment. In either case it is important to have a basic understanding of the technique and to know its pitfalls as well as its strengths. The techniques discussed here are used in the laboratory; however, as will be seen in the next section, many can and are used in the field.

**Fig. 6.2.** Polarization curves, corrosion potentials and corrosion rates in the presence of chlorides. (A) Anodic polarization curve in the presence of chlorides showing the pitting potential and the sharp increase in corrosion current above this potential (the dotted line represents the curve that was obtained in the absence of chlorides, as in Fig. 6.1(A)). (B) Influence of increasing chloride content on the anodic polarization curves. (C) Schematic presentation of the intersection of the cathodic polarization curves (as in Fig.6.1(B)) and the anodic polarization curves with increasing chloride content, showing the increase in corrosion rates and decrease in corrosion potential (i.e. becoming more negative) with increase in chloride content.

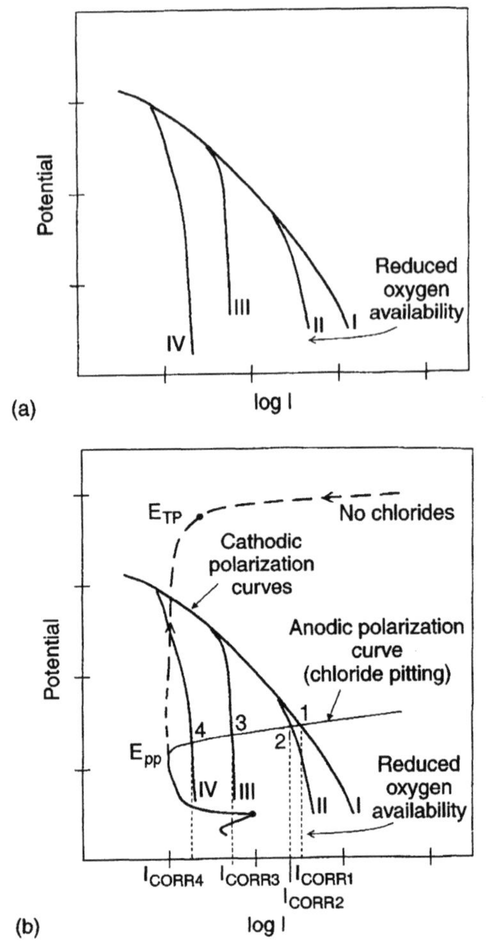


**Fig. 6.3.** Schematic presentation of the effect of availability of oxygen (dissolved in the solution) on the polarization curves, corrosion potential and corrosion rates in systems with chloride present. (A) Cathodic polarization curves affected by oxygen availability. (B) Intersection of the cathodic polarization curves with an anodic polarization curve of a system with chlorides, showing reduction in corrosion potential and in the corrosion rate in the system where oxygen is limited.

### 6.2.1. Visual Techniques and Mass Loss

Visual techniques involve the simple observation of whether or not the steel corrodes in a structure to observations of damage in a concrete specimen, to breaking open concrete or mortar to examine the bars, and finally to microscopic analysis. A detailed analysis is almost always destructive as the steel must be removed from its environment.

Typical methods include measurements of the fraction of the surface area corroded, number of pits per unit area, and pit depth. Details of the

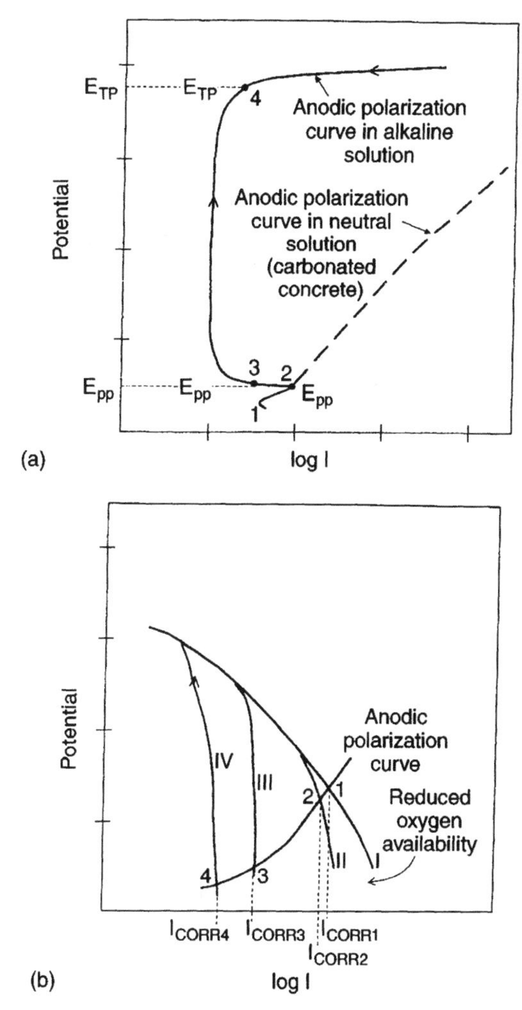


**Fig. 6.4.** Schematic presentation of the effect of carbonation and availability of oxygen on the polarization curves, corrosion potentials and corrosion rates in systems without chlorides. (A) Effect of carbonation on the anodic polarization curve showing the elimination of passivation (the solid line is the curve in passivating conditions, as in Fig. 6.1(A)). (B) The intersection of the anodic and cathodic polarization curves showing the reduction in corrosion potential (becoming more negative) and corrosion rates with reduction in oxygen availability.

procedures are given in several standards such as ASTM G46-94 (Standard Guide for Examination and Evaluation of Pitting Corrosion). It should be noted that if pitting corrosion has occurred for a significant length of time or spalling and cracking is present, the pits might have coalesced and what started as pitting corrosion might look like general corrosion.

Measurements of mass loss require the preweighing of bars before exposure. They also require assurance that cleaning and exposure techniques do not introduce mass loss unrelated to the corrosion processes being studied. In chloride-induced corrosion, which is a pitting phenomenon, mass loss are measurements of little value.

### 6.2.2. Electrochemical Techniques

Electrochemical techniques are extremely useful in assessing the corrosion behaviour of steel in concrete. Several of the techniques are non-destructive and can be used to follow the corrosion of specimens over time.

#### *6.2.2.1. Corrosion Potential*

The simplest technique is the measurement of the corrosion potential. As noted in section 6.1, the corrosion potential represents the potential which the rates of the anode and cathode processes are in equilibrium under a given condition. The potential can be assessed by simply measuring the voltage difference between a reference electrode and the steel. However, the actual measurements need to be carried out with attention paid to several points. First of all, reference electrodes must be properly maintained, and periodically compared to a primary reference electrode. If this is not done all of the results will be in question. A high-input impedance (>10 Megaohm) voltmeter is required. It should be capable of reading to 1 mV resolution, and a range up to 2 V. For small laboratory specimens, one usually places the reference electrode in a salt bridge (clean reservoir of solutions with a fretted or low liquid loss to prevent contamination) or on a wet sponge to measure the potential. If the steel is in concrete or mortar, then the cover must be moist entirely to the level of to the steel to obtain a stable reading. In making the measurement the steel is connected to the positive end of the voltmeter and the reference electrode to the ground. For further information see ASTM C 876- 91 (Test Method for Half Cell Potentials of Uncoated Reinforcing Steel in Concrete). A typical set-up is shown in Fig. 6.5.

#### *6.2.2.2. Polarization Resistance*

The polarization resistance technique is used to estimate the corrosion rate of steel in concrete or mortar. In order to understand the use of this method, it is first necessary to review some additional points from corrosion theory. At the corrosion potential, $E_{corr}$, the anodic and cathodic currents are equal in magnitude. The application of an external current source will move the potential away from $E_{corr}$. This is known as polarization. If the polarization is within several mV of $E_{corr}$, the potential change

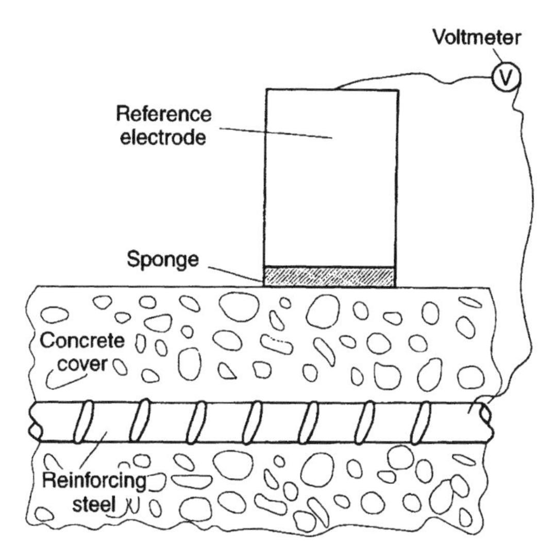


**Fig. 6.5.** Schematic description of potential measurements.

is almost always proportional to the current applied as shown in Fig. 6.6. Stern and Geary [6.1] showed that the slope of the voltage versus current curve at zero applied current was inversely proportional to the corrosion rate as shown below:

$$\Delta E/\Delta i|_{i=0} = R_p \text{ or } I_{corr} = B/R_p, \tag{6.1}$$

where i is the current density applied, $I_{corr}$ is the corrosion rate (current), $R_p$ is the polarization resistance, and B is a constant. Typical values for B are 26 mV. The units for $I_{corr}$ are typically μA/cm², and $R_p$ is in ohm•cm².

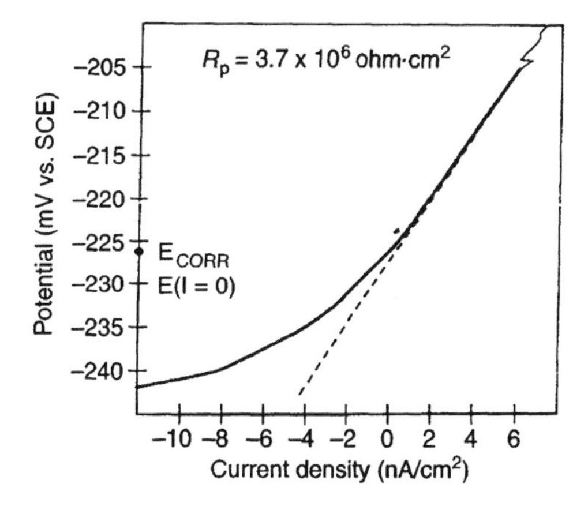


**Fig. 6.6.** Polarization resistance curve. (After Ref. 6.4)

This measurement technique is described in the ASTM G59-91 (Practice for Conducting Potentiodynamic Polarization Resistance Measurements), and requires a potentiostat and certain auxiliary items. A potentiostat is used to displace the equilibrium of the steel by about several mV and the current between the steel and the counter electrode is measured, as shown in Fig. 6.7. In practice, the potential is initially set about 20 mV below the corrosion potential, then swept at a low scan rate of about 0.1 mV/s to a potential of about 20 mV above $E_{corr}$. The current response during this scan is recorded. This is presented in Fig. 6.6, which shows the linear nature of the current versus voltage relationship. The polarization resistance $R_p$ is obtained from the slope of the current vs. voltage response at the point of zero current.

In general it has been found that when $R_p$ is >66 000 ohm•cm² (approximately 0.4 μA/cm²), the corrosion rate is low [6.2]. Note that the conversion to a current density assumed a particular value for the constant B. The constant B is not easily determined. In general, Stern and Geary [6.1] pointed out that the uncertainty inherent in this technique extends over a factor of 2 to 4.

A further complication when working in concrete systems is that the resistivity of the concrete, $R_{concrete}$, can be quite high. This high resistivity can lead to an erroneous result, as much as a factor of 2 or more away from the true $I_{corr}$. The correct polarization resistance is:

$$R_p = R_{pmeasured} - R_{concrete} \tag{6.2}$$

$R_{concrete}$ can either be independently determined, or by using a current interruption technique it can be automatically compensated for. The current interruption technique is available on many of the newer potentiostats. A complete discussion of these effects is given in reference [6.3].

#### *6.2.2.3. Electrochemical Impedance Spectroscopy (EIS)*

The EIS technique works on the same principles as the polarization resistance technique, with the difference being that an alternating potential is applied at varying frequencies, and the resulting alternating current is measured. The impedance or alternating voltage divided by the alternating current is plotted as a function of frequency. The basis for this technique is that steel in concrete behaves like a circuit as shown in Fig. 6.8. The $R_c$ resistor represents the concrete resistance and $R_p$ is the polarization resistance. The capacitance $C_{d.l.}$ is related to the double layer that forms in electrochemical systems at the metal/solution interface. The discussion of double layers is beyond the scope of this book. The equation governing the impedance response in the circuit shown in Fig. 6.8 is:

$$Z = R_c + R_p(1-j\omega CR_p)/[1+(\omega CR_p)^2] \tag{6.3}$$

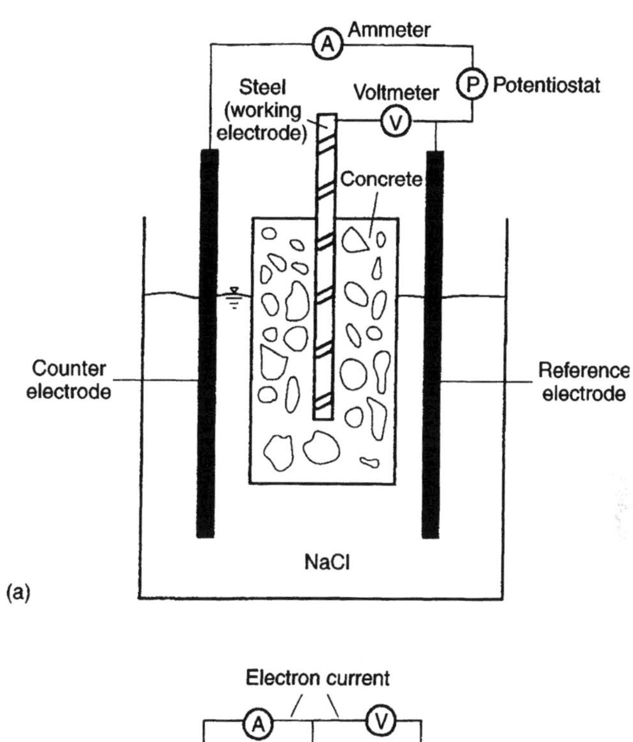


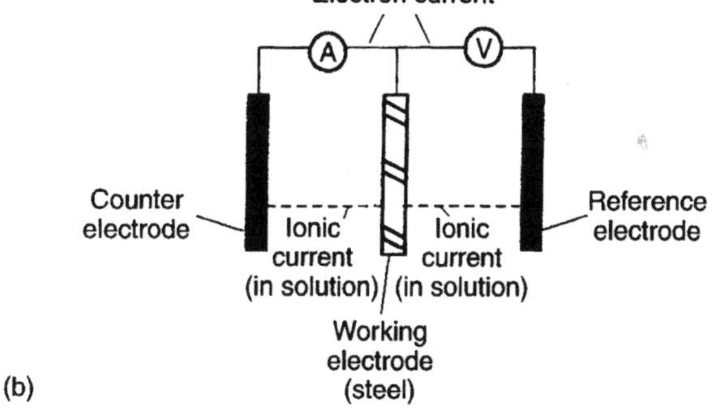


**Fig. 6.7.** Schematic description of polarization testing. (A) Schematic description of the test set- up. (B) Schematic description of the electrical circuit.

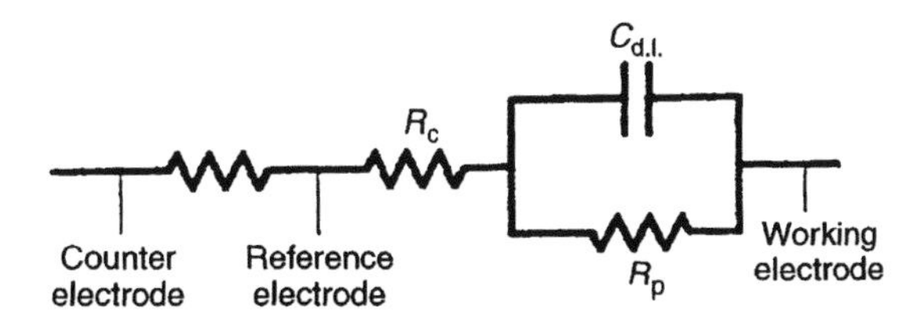


**Fig. 6.8.** Electrical circuit to model the response of concrete to polarization by external source. (After Ref. 6.4).

where Z is the total impedance measured at frequency f, Hz; $R_c$ is the electrical resistance of the concrete; $R_p$ is the polarization resistance; $\omega = 2\pi f$ rads/s; and $j = (-1)^{1/2}$.

From this equation it can be seen that when the frequency is approaching infinity, the impedance is equal to $R_c$. When the frequency approaches zero the impedance is essentially equal to $R_c + R_p$.

The results of the EIS test can be plotted as a Bode plot shown in Fig. 6.9. From this plot the values of $R_p$ and $R_c$ can be determined. The $I_{corr}$ can be calculated using eqn. (6.1). When used in the laboratory, the range of frequencies in which the test is conducted is from about 0.2 mHz to 40 000 Hz.

Additional information can be obtained by measuring the phase angle between the voltage and current. If capacitance effects are dominating then the phase angle approaches –90°. For further information there are several references that discuss the technique in greater detail [6.4, 6.5].

#### 6.2.2.4. *Cyclic Polarization*

This technique is useful in determining the pitting tendency of a metal in a particular environment. In addition, it is useful for demonstrating whether an inhibitor is an anodic inhibitor or a cathodic inhibitor. To simulate steel in concrete, a chloride solution that is saturated in calcium hydroxide is used. Optionally, sodium and potassium hydroxide may be added. Page and Treadaway [6.6] have shown that saturated calcium hydroxide actually gives similar results to those obtained for steel in cement pastes.

The test uses a potentiostat to shift the potential of the steel from its normal $E_{corr}$ level to much lower potentials of the order of approximately –800 mV versus SCE. The potential is then slowly altered in a positive direction at a rate slow enough for the formation of a protective film. The

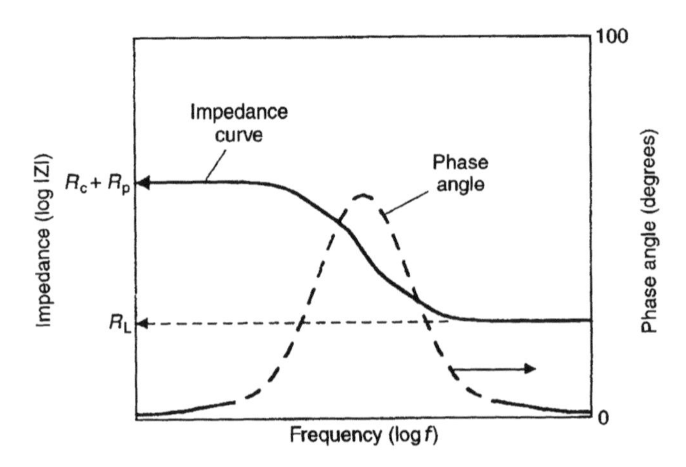


**Fig. 6.9.** Bode plot describing the realities of electrical impedance testing. (After Ref. 6.4).

scan continues in that direction until there is a sharp increase in the current, indicating that either pitting or transpassive zone has been reached (Fig. 6.10). When the current reaches a high preset value the scan is reversed. This latter scan is performed quickly so as to minimize the amount of time that the specimen is subjected to high current. Berke and Hicks [6.4] have shown that a constant scan in both directions at about rate of 5 mV/s provides good results. Conducting the tests in the aqueous solution avoids errors caused by the large resistance of concrete. This test is covered by ASTM G61 (Test Method for Conducting Cyclic Potentiodynamic Polarization Measurements for Localized Corrosion, Susceptibility of Iron-, Nickel-, or Cobalt-Based Alloys).

Because the applied potential brings the sample into regions far removed from the corrosion potential (500 mV or more away from $E_{corr}$), this technique is destructive. Under actual field conditions, such high values are not encountered unless large stray voltages are present. The destructive nature of the test makes it unsuitable for concrete structures in the field, nor is it a monitoring technique. Its use is restricted to the examination of test specimens in the laboratory.

Two potentials, namely the pitting or protection potential $E_p$ and the nucleation or breakdown potential, $E_b$, are of interest in examining Fig. 6.10. $E_p$ is the potential below which pitting cannot occur and is taken to be the potential at which the forward scan intersects the reverse scan. $E_b$ is the potential at which a pit first nucleates and occurs when the forward scan breaks away from the polarized curve in which there is no pitting. If pitting is not present, there is minimal hysteresis between the forward and reverse scans (dotted line in Fig. 6.10).

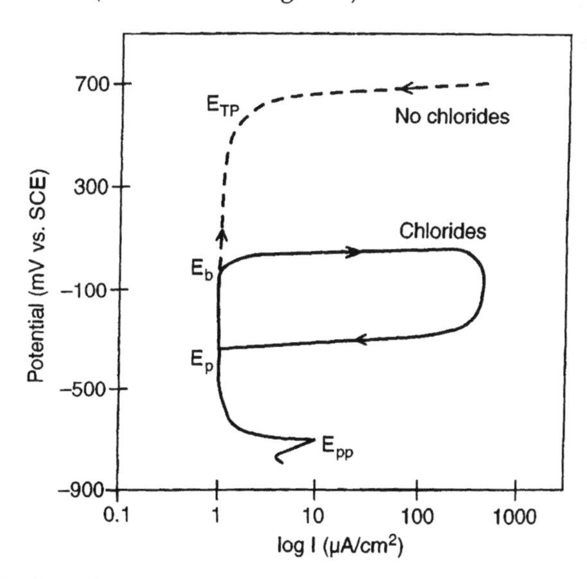


**Fig. 6.10.** Cyclic polarization curve of steel in de-aerated alkaline solution. (After Ref. 6.4).

*6.2.2.5. Macrocell Techniques*

These techniques involve the measurement of the current passing between discrete embedded metal members. Typically, one member (anodic or corroding reinforcing bar) is in a more severe exposure than the other which becomes the cathodic. In concrete this is most easily accomplished by embedding several steel bars at different depths and exposing the concrete to chloride or carbonation in a manner to obtain depassivation only of the steel destined to be the anode. Two standard methods of performing this test are the Southern Exposure Method [6.7] and ASTM G109-92 (Test Method for Determining the Effects of Chemical Admixtures on the Corrosion of Embedded Steel Reinforcement in Concrete Exposed to Chloride Environments).

Another configuration involves the use of an 'inert cathode' such as stainless steel or graphite. This allows for a smaller specimen as the members can be at the same depth. Since the cathode does not corrode it can be in the same exposure as the anode. This permits closer placement which minimizes resistive effects of the concrete.

The macrocell current is defined as the current between the corroding and cathodic members, and a measurement of an appreciable macrocell is often a good indication that corrosion has initiated. However, this is a galvanic current, and does not represent the localized microcell corrosion sites on the corroding steel. Also, in some cases the supposedly inert bar may go into corrosion, the result of which may be appearance of a current with the reverse sign of what is expected. In other cases where the resistivity is very high macrocells might not occur.

Even though most corrosion engineers realize that the macrocell current measures only a fraction of the total corrosion activity, the technique is often used in practice because of the simplicity of the measuring process. Figure 6.11 gives a schematic of a macrocell specimen as used in ASTM G109-92 (Test Method for Determining the Effects of Chemical Admixtures on the Corrosion of Embedded Steel Reinforcement in Concrete Exposed to Chloride Environments). The macrocell current is determined by simply measuring the voltage across the resistor connecting the top bar to the two bottom cathodic bars. The equation is:

$$I = V/R \tag{6.4}$$

where $I$ is the macrocell current (microamperes), $V$ is the voltage across the resistor (mV), and $R$ is the resistance of the resistor ($\Omega$). In other methods, a zero-impedance ammeter is connected between the cathodic and anodic bars to measure the current directly. If a resistor is used its resistance should be less than the ionic resistance between the bars so as to prevent the elimination of the macrocell by resistance that is too high.

Macrocell current measurements are not destructive, so macrocell currents can be determined as a function of time for the same specimen.

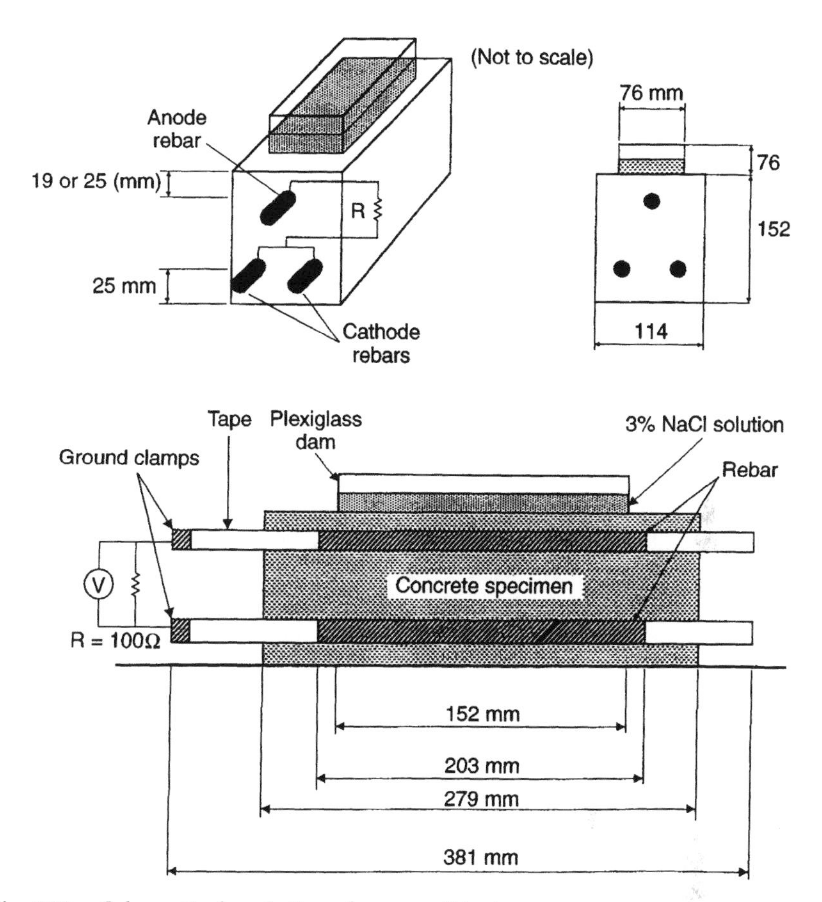


**Fig. 6.11.** Schematic description of macrocell test.

Often the current is integrated and the total charge passed is plotted as a function of time.

### *6.2.2.6. Potentiostatic/Galvanostatic Measurements*

These tests involve polarizing the metal to a constant potential and monitoring the current necessary to maintain the potential, or applying a constant current and monitoring the potential. Under potentiostatic conditions a sharp increase in current is indicative of corrosion. In the galvanostatic test a sharp decrease in potential to negative values is indicative of corrosion activities.

These tests can be destructive since they usually polarize the specimen to a level more than 100 mV above the corrosion potential. Their main uses are to determine if concrete admixtures are corrosive to embedded steel, or to determine if inhibitors are effective against admixed chlorides.

Potentiostatic procedures have also been used to accelerate chloride ingress since the negative chloride ions can be driven towards the bar if it is anodic. In this case differences in concrete resistivity can produce varying results as the actual potential at the steel can be significantly different than the potential measured. This difference is in addition to any differences in chloride ingress. For example, a lower-resistivity concrete might have a greater chloride content at the reinforcing bar after a given time than a higher-resistivity concrete. Furthermore, effects due to the concrete resistance are higher in the higher-resistivity concrete so the polarization is less, thus making it less severe. A correction for $R_c$ will prevent the latter problem and the reduced chloride ingress might be considered to be a property determined by the test method. In the normal application of the test method, the specified procedures of ASTM C1202- 94 (Test Method for Electrical Indication of Concretes' Ability to Resist Chloride Ion Penetration) or AASHTO T277-89 (Standard Method of Test for Rapid Determination of the Chloride Permeability of Concrete) call for integrating the current over a period of 6 hours. It is presumed that this integrated current, in units of Coulombs, is related to the permeability of the concrete to chloride ions. The validity of this has been questioned on various grounds. Additional details of the apparatus used in this test are given in section 6.2.2.7.

When used to determine the corrosivity of admixtures or the effectiveness of an inhibitor to admixed chloride, the test method is usually carried out with mixes of cement low in $C_3A$ content (<5%). This minimizes binding of chloride and other potentially harmful chemicals in the candidate admixture. In one technique, small mortar cylinders with an embedded reinforcing bar are cured for 3 days in a sealed mould, immersed in lime water for 24 hours and polarized at either +100 mV versus SCE (for inhibitor evaluations) or +260 mV versus SCE for 24 hours. A non-corrosive admixture will have an average current density of <1 μA/cm². An anodic corrosion inhibitor should have an efficiency >90% where efficiency is defined by: Efficiency = $[1-I_{inhibitor}/I_{control}]\times 100\%$, where $I_{inhibitor}$ is the current measured with the inhibitor present and $I_{control}$ is the average current with the chloride alone.

### 6.2.2.7. Conductivity Measurements

The conductivity of concrete is related to its permeability, and it also gives an indication of the ease at which macrocell corrosion can proceed. The inverse of conductivity is resistivity; the latter is actually what is determined in many of the test methods.

Concrete has relatively high resistivity (>3 kΩ•cm) and sometimes in the range of several hundreds of kΩ•cm. Thus, testing of resistivity requires a large DC voltage or lower AC voltage in the determination of the resistivity or conductivity. As noted in section 6.2.2.3, EIS techniques at high frequency ($\approx$20 000 Hz) can be used to determine the resistivity. Typically, a concrete lollipop is used in which a steel rod is embedded in a concrete cylinder. The lollipop is immersed in salt water which has a resistivity of about 1 Ω•cm. Thus, the solution boundary with the concrete can be considered to be the surface of a conductive cylinder and the steel bar's outside surface is the other cylinder with the concrete being the material of unknown resistance. A schematic of the set-up is given in Fig. 6.12. In that case the resistivity can be determined from the measured resistance using the following equation:

$$R = \{[\ln(a/b)]\rho\}/2\pi l \tag{6.5}$$

where $R$ is resistance ($R_{concrete}$/reinforcing bar surface area); $a$ is radius of the reinforcing bar; $b$ is radius of the concrete; $l$ is length of the bar; and $\rho$ is resistivity.

From eqn. (6.5) it can be seen that when a 9.5-mm diameter bar (#3 reinforcing bar) is used with a 3-inch (75-mm) diameter concrete cylinder, the resistivity ($\rho$) is fortuitously about equal to the ohmic resistance $R_c$ (defined as $R_{concrete}$ in eqn. (6.5)). $R_c$ is the high-frequency impedance value shown in Fig. 6.9. Measurement of $R_{concrete}$ is a non-destructive test method. Increases in resistivity over time are frequently noted [6.2, 6.8, 6.9], and it indicates that the concrete is becoming less permeable [6.8–6.10]. Correspondingly, a sharp decrease in $R_{concrete}$ is often associated with the onset of severe corrosion and perhaps the development of microcracking [6.2].

Another procedure for determining resistivity or conductivity can be carried out by the ASTM C1202 apparatus (ASTM C1202-94: Test Method for Electrical Indication of Concretes' Ability to Resist Chloride Ion

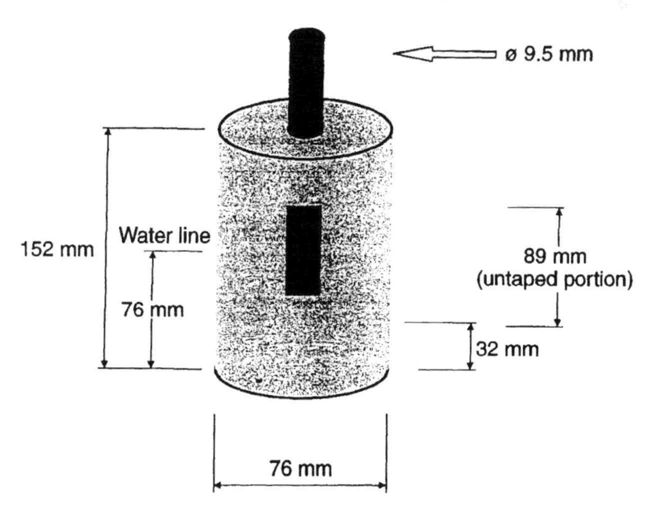


**Fig. 6.12.** Schematic description of 'lollipop' test for determining resistivity or conductivity. The electrical measurement circuit is shown schematically in Fig. 6.7.

Penetration) which is shown schematically in Fig. 6.13. The test was developed to measure a parameter related to the permeability of the system to chlorides. It involves placing a 2-in (50-mm) thick by 4-in (100-mm) diameter concrete disk in a cell in which a sodium chloride solution is contained in a reservoir on one side, and a sodium hydroxide solution on the other. A 60-V DC potential is applied, and the resistivity current is measured as shown in Fig. 6.13. Resisitivity can be determined from the current reading since the resistance of the disk can be calculated immediately from Ohm's Law:

$$R=V/I \tag{6.6}$$

where $R$ is resistance; $V$ is voltage; and $I$ is current.

The resistivity ($\rho$) is determined from:

$$\rho=RA/l \tag{6.7}$$

where $A$ is area of the disk; and $l$ is thickness of the disk.

Excellent correlations between $\rho$ determined by the EIS technique and the $\rho$ determined from the initial current reading in the ASTM C1202-94 method have been noted [6.8, 6.11]. Thus, although this method was developed as a means for estimating the permeability to chlorides, it provides a direct measurement of electrical resistivity. The correlation between the electrical resistivity and chloride permeability is not always straight forward.

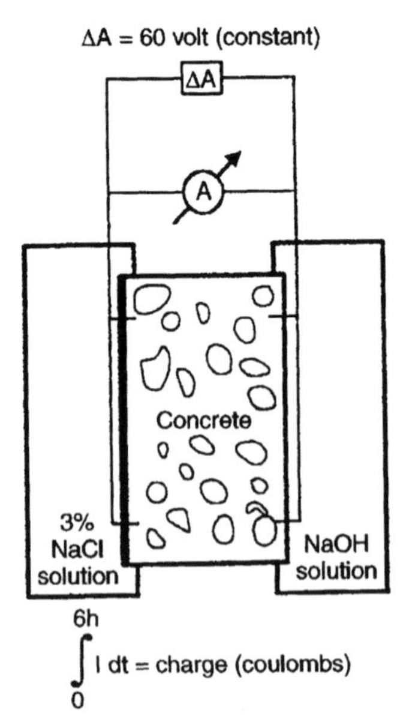


**Fig. 6.13.** Schematic description of the ASTM C1202-94 test method for electrical indication of concrete's ability to resist chloride ion penetration (known also as rapid chloride permeability test).

## 6.3. FIELD MEASUREMENTS

Assessment of the extent of corrosion and ongoing corrosion activity in field structures is much more difficult than in laboratory specimens. In field structures, well-defined specimen configurations, controlled exposure conditions, and easy access to the embedded steel reinforcement do not exist. Other problems often include a lack of information about mixture proportions, and reinforcement size and locations. The presence of stray voltages, changing weather conditions, and other interferences to electrochemical measurements frequently complicate the interpretation of measured results. In this section several methods of assessing corrosion activity are addressed. The cumulative use of these techniques is often required to overcome the difficulties mentioned above.

### 6.3.1. Visual Inspection and Delamination Survey

One of the best and least difficult forms of inspection is visually to assess the condition of the structure. One looks for signs of rust staining, concrete spalling and excessive cracking. If all of these are present the structure is most likely suffering some form of distress, likely corrosion-induced. Exposed reinforcing bars are sure signs that corrosion is occurring. However, rust stains alone sometimes come from exterior sources (such as a rusting object that was located on or near the concrete).

It is also necessary to test for delamination. Severe corrosion activity frequently results in the delamination of the concrete at the level of the upper reinforcement (see Fig. 5.1). Concrete may look sound, but if delamination has occurred it will not be bonded to the concrete below. The chain drag method is commonly used to test for such delamination. When the chain being dragged passes over the delaminated region, there is a noticeable change in the sound of the chain on the concrete. Alternatively, a heavy mallet can also be used for delamination identification being detected by the changed sound produced on impact. This technique is quite accurate, but requires an experienced operator. Also, at noisy locations it can be difficult to resolve the differences in the sound produced between delamination and the intact regions of the concrete.

More sophisticated techniques for detecting delamination include radar, X-ray, and ultrasonic pulses. Such techniques are quite expensive, and at the present time are not necessarily superior to the chain-drag technique.

If the visual survey and indication of delamination suggest that corrosion is occurring, additional assessment techniques as noted in the following sections should be carried out to determine if corrosion has just begun, but is not yet at an advanced enough stage to cause visual distress in other parts of the structure. However, even if the surveys show that there are no visible signs of corrosion or delamination, the additional tests might be desirable to develop baseline information for future maintenance operations.

### 6.3.2. Potential Mapping

Potential maps provide useful tools for determining if there is a high probability of corrosion activity. The basic technique is as discussed in section 6.2.2.1, and is conducted according to ASTM C876-91 (Test Method for Half Cell Potentials of Uncoated Reinforcing Steel in Concrete). Preparation of the structure is important and needs additional discussion, as does interpretation of the results.

For the corrosion potentials to be accurately measured one needs to ensure that the concrete between the surface and the reinforcement be sufficiently wet to be electrically conductive. This will often require that the structure be ponded or sprayed with water the night before, and that measurements be taken in the early morning before the sun dries the surface. Any puddles of water on the surface to be measured need to be removed to localize the area being measured. It is always necessary to test for electrical continuity between the reinforcing bars. This is performed by drilling holes to different bars and measuring the resistance between them. If the resistance is above 1 Ω, then continuity may not be present. In that case one might have to locate bars individually to perform the test on each one separately. This seldom occurs unless epoxy-coated steel is present. It should be emphasized as discussed in sections 2.3 and 6.1 that the corrosion potential may not itself be an indicator of the rate of corrosion.

It is generally more useful to produce a corrosion potential contour map than to rely on absolute corrosion potential values at scattered locations on the structure. The potential readings are taken on a grid with a typical spacing of about 1 m. A closer spacing will significantly improve resolution and the ability to identify potentially corroding sections; conversely, higher spacings will reduce the certainty of finding all of the corroding areas.

An example of a potential contour map is given in Fig. 6.14 for a bridge deck that has been exposed to de-icing salts. Typically, those areas that have more negative corrosion potentials than adjacent regions should be identified. When chloride is present corrosion is most likely in the regions where the corrosion potential contour lines are most dense, as in the bottom left-hand corner of Fig. 6.14.

Once again, it should be noted that if the potential map indicates that corrosion is likely in a given area, its presence still needs to be verified by other test methods. If the potential mapping indicates a low probability of corrosion activity, the additional testing required might be limited.

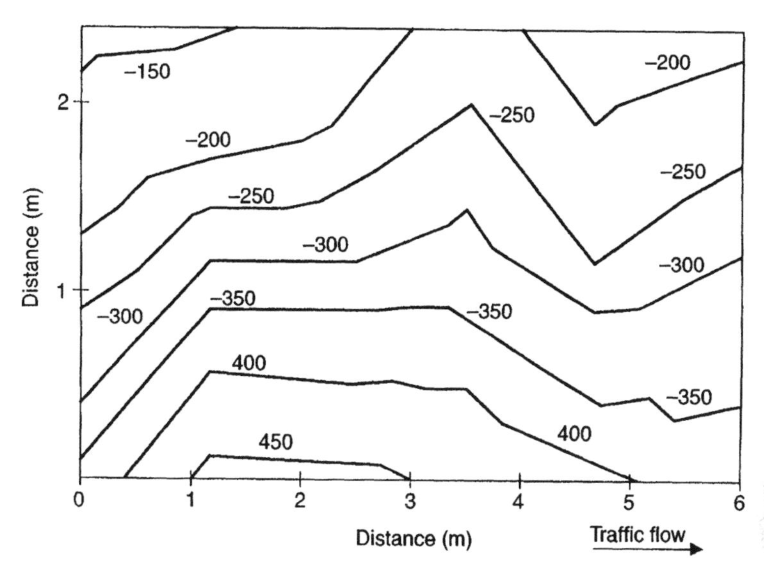


**Fig. 6.14.** Example of potential map of bridge deck (values are mV vs. SCE).

### 6.3.3. Confirmation of Potential Mapping

Since potential measurements are not uniquely related to the corrosion rate of steel in concrete, the additional techniques described below are required to determine the actual extents and rates of corrosion in a given area.

#### *6.3.3.1. Detailed Visual Inspection*

This is similar in nature to the initial visual inspection discussed in section 6.3.1. The main difference is that the areas indicated as having the most negative potentials in the corrosion potential contour plots are carefully examined. An additional delamination survey might be carried out in the suspected area and in adjacent areas. If the area has already been identified as being one with staining, delamination, spalling or exposed reinforcement, the potential contour map is seen as probably representing the state of corrosion in these areas.

#### *6.3.3.2. Chloride and Carbonation Analysis*

Chloride analyses of the concrete at the level of the reinforcing steel are useful in determining if there is sufficient chloride present to cause depassivation of the steel. As noted earlier, chloride levels of 0.9 to 1.2 $kg/m^3$ of concrete are considered to be the threshold level for corrosion initiation. If chloride analyses at the reinforcement show values above this threshold in regions identified as suspect in the corrosion potential contour map, there is a strong probability of corrosion being present, even with the absence of visual evidence. Note that the recent Strategic Highway Research Program (SHRP) report [6.12] pointed out that there is about a 5-year time lag between corrosion initiation and visible evidence being observable on the surfaces of bridge decks.

It is useful not only to measure chloride at the level of the steel, but also to measure the chloride as a function of depth. If such measurements are performed and the age of the structure known, the data can be fitted to a diffusion equation. The effective diffusion coefficient, and information on the level of chloride at the surface can then be used to predict future chloride profiles. These profiles become a useful tool in predicting future corrosion activity by providing indications of the additional time expected before corrosion will be initiated. The ingress of chloride is relatively slow in most field structures. Consequently, this method of calculation is applicable only to structures that have been in service for at least 3–5 years in severe exposures or 10 or more years in a less severe exposure. A minimum of measurements of at least three different depth levels is needed, and it is preferable to have more. Sampling should be carried out in increments of 12.5 mm or less, and be spaced from the surface to below the outer reinforcing level.

If the presence of chlorides in the surrounding environment is unlikely, or chloride is not found in the above analyses, tests for carbonation at the reinforcing bar level should be carried out. This can be achieved by examination of cores with phenolphthalein solution (see section 4.3.1.4).The application of the solution should be made immediately after coring on site. Delay in the application of the solution may lead to erroneous results, as the exposed core surface may be carbonated during the handling after it had been cored.

#### *6.3.3.3. Corrosion Rate Measurements*

Further confirmation of corrosion activity is possible using corrosion rate measurements. Such measurements ordinarily involve measuring the polarization resistance. As noted earlier, the resistivity of concrete is appreciable; this can cause the corrosion rate to be severely underestimated. Fortunately, almost all commercial instruments now have methods to correct for this.

Laboratory experiments have indicated that to identify the onset of corrosion using EIS, it is necessary to employ frequencies as low as 1 mHz as the lowest frequency [6.4]. This allows one to clearly determine whether an inflection point in the Bode diagram is occurring at or above 10 mHz (see Fig. 6.9). The occurrence of such an inflection point is indicative of corrosion. However, such measurement requires a measurement time of more than 20 minutes at each location. Not only is this impractical from the point of view of the number of positions that should be assessed, it is also very difficult to assume that conditions will remain constant in the field for 20 to 30 minutes. Therefore, polarization resistance is more useful then EIS as a field measurement.

It is necessary to decide on a scan rate for polarization resistance measurement to be made. At the early corrosion stage, the EIS data suggest that the scan rate should not exceed about 0.2 mV/s. Higher scan rates will only identify regions that are undergoing rapid corrosion and could be very inaccurate in regions where corrosion is being initiated. Since highly corroding areas are usually readily identified by visual observations, coupled with corrosion potential contour maps and are usually markedly high in chloride levels, it is of little additional benefit to identify them by polarization resistance measurements. It is of more interest to use such methods to identify areas that are in the initial stages of corrosion and that will become problem areas in the next 5 years. This requires the use of polarization resistance measurement carried out at low scan rates.

Figure 6.15 shows some corrosion rate measurements superimposed upon the corrosion contour map in Fig. 6.14. The values in Fig. 6.15 are point measurements of corrosion rates. They are expressed in units of Siemens (S), which is the inverse of the polarization resistance value $R_p$. The corrosion rates are indeed highest in the region of the left corner as expected. Of interest are the values in the 20 μS/cm² range, which is equivalent to about 6 μm/year corrosion rate. Such values are in the initiation range [6.2]. If a patch repair is performed omitting regions marked by corrosion rates in this range, they could be showing signs of premature distress in subsequent years.

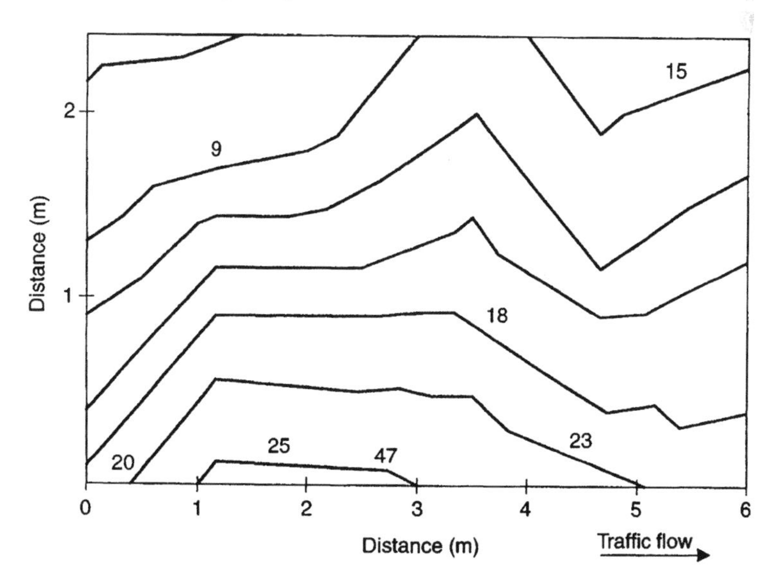


**Fig. 6.15.** Corrosion rate measurements (numerical values in μS/cm²) superimposed on the corrosion contour map of Fig. 6.14.

One problem in corrosion rate measurements is that of determining the area of steel that is being polarized. Using a relatively large counter electrode reduces the relative influence of this uncertainty. New guard ring devices to confine the current have been developed [6.13, 6.14] and are occasionally used. Different devices often give different corrosion rate values; however, high values with one device typically give high values with another. Discrepancies are due to using different constants in the corrosion equations, errors in determining the area polarized, and to using scan rates that are too rapid as noted above.

#### *6.3.3.4. Analysis of Cores*

The last method of field analysis is the removal of relatively large sections of concrete, often through the reinforcement. This typically involves using a core drill to obtain 100-mm or larger diameter samples that are about 100 mm long. These specimens can visually confirm the condition of reinforcing steel and the concrete quality. Carbonation can be identified if present and detailed chloride analyses can be performed. If major rehabilitation is planned, the taking of cores to back up the decision is recommended.

#### *6.3.3.5. Assessment of Epoxy-Coated Reinforcing Bars*

Epoxy-coated reinforcing steel presents a problem because of its lack of electrical continuity. As noted earlier this makes the process of performing electrochemical measurements more tedious as more connections are needed. Another problem is that corrosion, when present, is occurring at small defects in the coating or under the coating and as such is hard to readily identify. One method to identify such pinhole defects (sometimes called 'holidays') is to use EIS at a high frequency (1000 Hz or higher). The counter electrode and reference electrode is moved down a bar and a drop in the resistance is indicative of a pinhole defect. This can be coupled with coring and chloride analysis to obtain some idea as to the performance of the steel.

## REFERENCES

6.1. Stearn, M.S. & Geary, A.J., Electrochemical Polarization I: A Theoretical Analysis of the Slope of Polarization Curve. *Journal of the Electrochemical Society*, **104**(1) (1957), 56–63.
6.2. Berke, N.S., The Effects of Calcium Nitrite and Mix Design on the Corrosion Resistance of Steel in Concrete, Part 2: Long Term Results. In *Corrosion 87, The NACE Annual Conference and Corrosion Show*, National Association of Corrosion Engineers, Houston, Texas, 1987, paper 132.
6.3. Berke, N.S., Shen, D.F. & Sundberg, K.M., Comparison of Current Interruption and Electrochemical Impedance Techniques in the

Determination of the Corrosion Rates of Steel in Concrete. In *The Measurement and Correction of Electrolyte Resistance in Electrochemical Tests*, eds L.L. Scrinber and S.R. Taylor. ASTM STP 1056, American Society of Testing and Materials, Philadelphia, 1990, pp. 191–201.
6.4. Berke, N.S & Hicks, M.C., Electrochemical Methods of Determining the Corrosivity of Steel in Concrete. In ASTM STP 1000, 25th Anniversary Symposium Committee G1, American Society of Testing and Materials, Philadelphia, 1990, pp. 425–40.
6.5. Sagues, A.A., Critical Issues in Electrochemical Corrosion Measurement Techniques for Steel in Concrete. In *CORROSION 91*, The NACE Annual Conference and Corrosion Show, National Association of Corrosion Engineers, Houston, Texas, 1991, paper 141.
6.6. Page, C.L. & Treadaway, K.W.T., Aspects of the Electrochemistry of Steel in Concrete. *Nature*, **297** (1992), 109–15.
6.7. Pfeifer, D.W., Landgren, J.R., LeClaire, P.J. & Zoob, A., *Protective Systems for New Prestressed and Substructure Concrete – Pilot Time to Corrosion Studies*. Technical Report, Wiss, Janney, Elstner Associates, Inc., sponsored by the Federal Highway Administration, 1984.
6.8. Scali, M.J., Chin, D. & Berke, N.S., Effect of Microsilica and Fly-ash Upon the Microstructure and Permeability of Concrete. In *Proc. 9th Int. Conf. on Cement Microscopy*, International Cement Microscopy Association, Texas, 1987, pp. 375–9.
6.9. Berke, N.S., Scali, M.J., Regan, J.C. & Shen, D.F., Long Term Corrosion Resistance of Steel in Silica Fume and/or Fly-ash Concretes. In *Proc. 2nd CANMET/ACI Conference on Durability of Concrete*, ed. V.M. Malhotra. ACI SP-119 American Concrete Institute, Detroit, 1991, pp. 393–422.
6.10. Bamforth, P.B. & Chapman-Andrews, J.F., Long Term Performance of R.C. Elements Under UK Coastal Exposure Conditions. International Conference on Corrosion and Corrosion Protection of Steel in Concrete, University of Sheffield, Sheffield Academic Press, UK, 1994.
6.11. Berke, N.S. & Roberts, L.R., The Use of Concrete Admixtures to Provide Long Term Durability From Steel Corrosion. In *Proc. 3rd CANMET/ACI International Conference on Superplasticizers and Other Chemical Admixtures in Concrete*, ed. V.M. Malhotra. ACI SP-119, American Concrete Institute, Detroit, 1989, pp. 383–404.
6.12. Ballinger, C.A., Methodology for the Protection and Rehabilitation of Existing Reinforced Concrete Bridges. Quarterly Report No. 3, Wilbur Smith Associates, BTML Division, Report to Strategic Highway Research Program, SHRP-89-C104, 1991.
6.13. Feliu, S., Gonzales, J.A., Escudero, M.L., Feliu, S. & Andrade, M.C., Possibility of the Guard Ring for the Confinement of the Electrical Signal in Polarization Measurements of Reinforcements. In *CORROSION 89*, The NACE Annual Conference and Corrosion Show, National Association of Corrosion Engineers, Houston, Texas, 1989, paper 623.
6.14. Flis, J. Sehgal, A. & Li, D., Condition of Evaluation of Concrete Bridges Relative to Reinforcement Corrosion – Vol. 2, Method of Measuring Corrosion Rate of Reinforcing Steel', Strategic Highway Research Program, National Research Council, Final Report C-101, SHRP-S/FR-92-104, Washington, DC, 1992.

CHAPTER 7

# Corrosion Control

## 7.1. INTRODUCTION

The processes leading to corrosion of steel in concrete were discussed in Chapter 4. An understanding of these processes is essential for design of reinforced concrete structures that will have the characteristics required for corrosion control.

In most instances the corrosion control methods used can be described as passive, i.e. the durability performance is obtained by proper design and control of the concrete cover. Such means are usually specified in design codes: minimum concrete cover thickness, the inherent concrete properties (in terms either of design strength or maximum w/c ratio) and the maximum allowable crack width permitted to be induced by loading or environmental effects. The requirements in the various codes are based to a large extent on experience and on test results of both penetration of chlorides into concretes, and of the depth of carbonation developed over time. These will be discussed in sections 7.2 and 7.3.

When exposure conditions are particularly harsh, there is frequently a need to apply special measures, beyond the minimum called for by the codes. Those include both passive and active measures. Passive measures may include the specification of high-quality concretes produced by incorporation of various types of chemical admixtures (e.g. superplasticizers) and mineral admixtures (e.g. fly-ash, silica fume, slag). On top of that, the addition of corrosion inhibitors (special admixtures that can improve the passivation stability of the steel) may be specified. The use of coated reinforcing bars (epoxy-coated and galvanized reinforcing bars) has become popular. Finally, sealers applied to the finished concrete surface may be required for additional protection. An active corrosion protection system, such as catholic protection, may be specified as an alternative or may be implemented later on. These procedures are covered in section 7.4.

It should be borne in mind that the application of the various special means to enhance the corrosion resistance of steel in concrete requires caution. The special means are not intended to replace the standard requirements in the specifications, but should rather be additional to them. These special means can sometimes be unsuccessful or even detrimental if not properly designed and applied. It is essential for the engineer to understand the basic concepts underlying these means and their pitfalls, in order to use them adequately. These aspects will be included in the discussion in section 7.4.

## 7.2. CONTROL OF CARBONATION

### 7.2.1. Influences of Environmental Conditions

In discussions of the carbonation of concretes, it is essential to keep in mind that, as shown, in Fig. 7.1, the carbonation process is extremely sensitive to the moisture content of the concrete. At very high humidities the pores in the concrete will be almost entirely filled with water, and gaseous $CO_2$ will not penetrate readily. Conversely, at very low humidities, there is lack of essential moisture for the carbonation reaction to proceed.

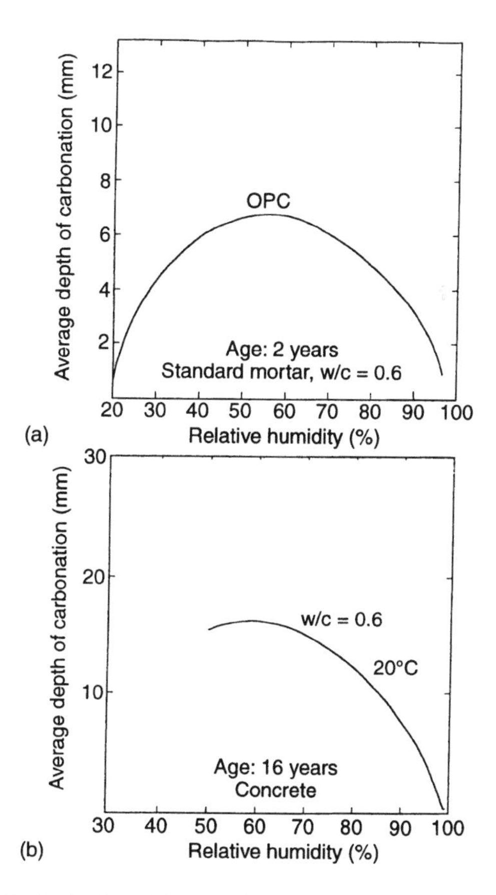


**Fig. 7.1.** Effect of relative humidity on the depth of carbonation at the ages of 2 and 16 years in 0.6 w/c ratio mortar and concrete. (Adapted from Ref. 7.1)

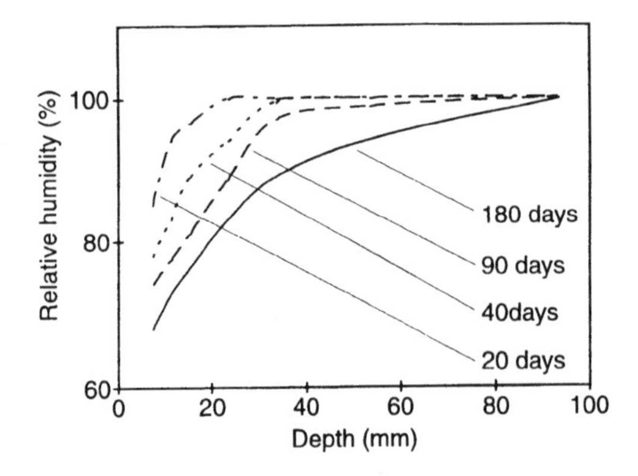


**Fig. 7.2.** Moisture gradients in 0.59 w/c ratio concrete dried from its surface in the laboratory at 60% relative humidity for various time periods. (Adapted from Ref. 7.2)

It should be borne in mind that the external relative humidity may be substantially different from the relative humidity in the pores of the concrete, since in practice concrete is rarely in equilibrium with its surroundings throughout all its depth. Concrete tends to dry slowly and moisture gradients such as those illustrated in Fig. 7.2 can be produced within the concrete cover. When assessing test results, or estimating rates of carbonation based on various models, it is essential to be aware of such effects which are difficult to quantify accurately.

Furthermore, one should consider not only the general conditions of the surrounding environment but also some 'microclimate' effects as sheltering from rain, as was illustrated in Fig. 4.17. In many instances, carbonation effect tests carried out in the laboratory involve carbonating under conditions of 40 to 60% RH, the range of humidity under which carbonation occurs most rapidly (Fig. 7.1). Therefore such laboratory tests constitute a worst-case scenario, as seen for example in the data in Fig. 7.3.

An important environmental factor is the ambient temperature. In common with most chemical reactions, the carbonation reaction is accelerated by increased temperature. The effect of temperature is demonstrated in Fig. 7.4 for 0.5 w/c ratio concrete after 15 months of exposure. Increasing the ambient temperature from 20 to 30°C increases the depth of carbonation by 50 to 100%.

### 7.2.2. Concrete Composition

The most important parameter controlling the rate of carbonation is the quality of the concrete concerned. This is a function of the composition of the binder (i.e. whether Portland cement or blended cement was used), the water/binder ratio, w/b, and the curing conditions. All three of these para-

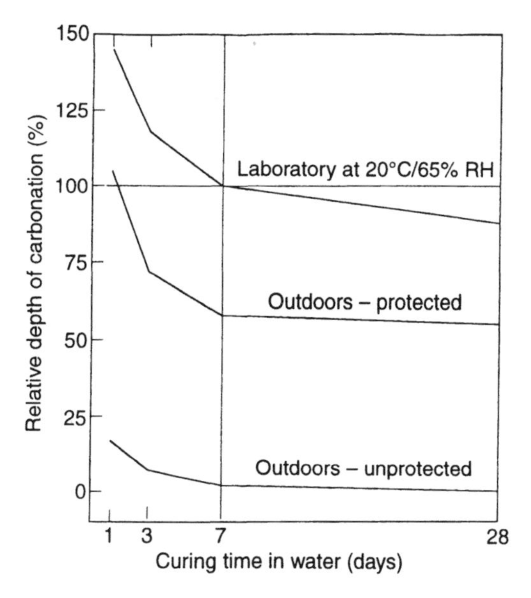


**Fig. 7.3.** Effect of curing and exposure conditions on the relative depth of carbonation of concrete at the age of 16 years. (Adapted from Ref. 7.1)

meters are directly within the scope of influence of the engineer, in the design and in the construction stages. It is thus essential that design engineers appreciate the extent of influence that these parameters will have.

Generally, denser concrete of low w/c ratio is expected to carbonate less, as suggested by the data in Fig. 7.4. There is an approximately linear relation between depth of carbonation and the w/c (or water/binder, w/b) ratio. A relationship of this kind was also reported by Hobbs [7.4] for both Portland cement and blended cement concretes (Fig. 7.5). The depth of carbonation was greater for the blended cements than for the Portland

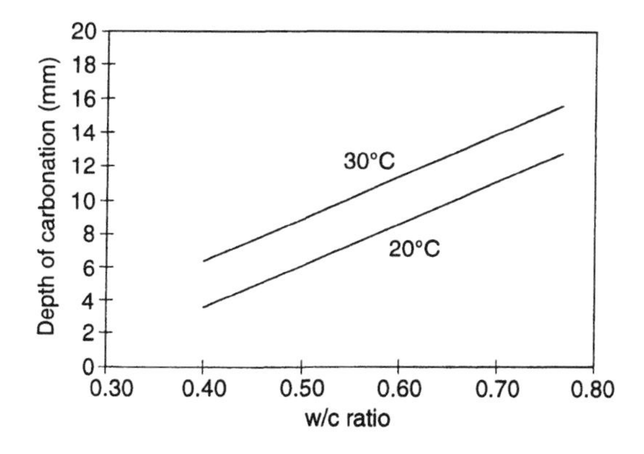


**Fig. 7.4.** Effect of temperature and w/c ratio on the depth of carbonation of concretes at the age of 15 months. (Adapted from Ref. 7.3)

cement concrete when compared on the basis of equal w/b ratios (Fig. 7.5). This is in agreement with other reports showing higher corrosion rates in fly-ash blended cement concretes, compared with Portland cement concretes of the same w/b ratio (Fig. 7.6). A trend of this kind was also observed for blended cements using blast furnace slag [7.6]. However, when compared on the basis of strength, the carbonation depth of concretes prepared with different cements seem to follow a common relation (Fig. 7.7). Results of this kind form the basis for specifying the concrete requirements in terms of strength.

It should be pointed out that specifications are not always given in terms of the strength grading of the concrete, but often are based on requirement of maximum w/c ratio. This is the case with the European pre-standard ENV 206. Bakker and Mathews [7.9] evaluated this standard by testing the carbonation depth of concretes of the same w/c ratio, consisting of Portland cement and blended cements complying with the European pre-standard ENV 197-1. They concluded that the carbonation depth of these concretes varied over a wide range, and suggested that it would be preferable to specify the concretes on the basis of strength grade, which correlated with carbonation depth, almost independently of the type of binder used. This suggestion is in agreement with the data presented above.

An alternative approach which is sometimes used when blended cement are applied is to define an efficiency coefficient, k, (between 0 and 1) for the fly-ash. An 'effective' w/c ratio is calculated, in which the content of the binder is not taken as its actual weight but rather as the product of the weight and the efficiency factor. Thus, the effective w/b ratio is taken

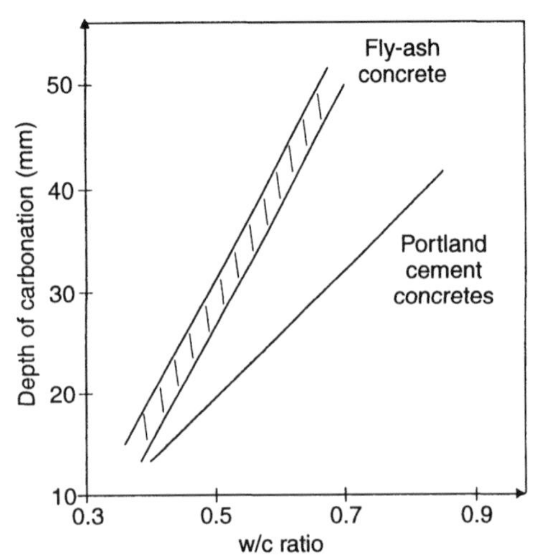


**Fig. 7.5.** Relations between w/b ratio and the depth of carbonation after 8.3 years, of Portland cement concretes and concretes with 35% fly-ash by weight of cement. (Adapted from Ref. 7.4)

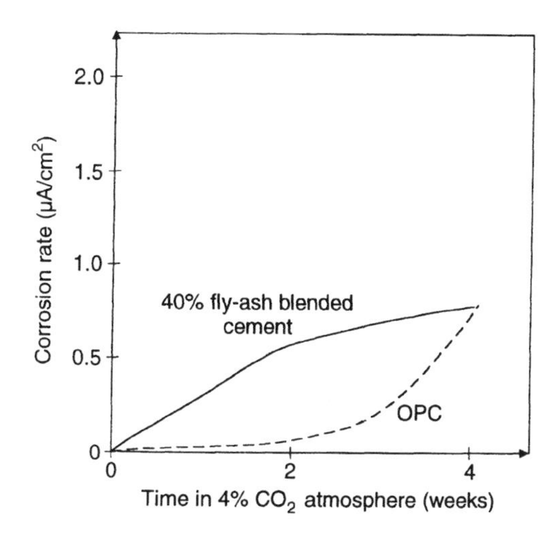


**Fig. 7.6.** Effect of fly-ash on the rates of corrosion of concretes of similar w/b ratio exposed to accelerated carbonation. (Adapted from Ref. 7.5)

as w/(c+k•p), where p is the content of the mineral admixture (e.g. fly-ash). Replotting the data in Fig. 7.5 using an efficiency factor of 0.3 yields a linear relation (Fig. 7.8) which is similar to the linear relation obtained for concrete with Portland cement (Fig. 7.5). It should be noted that in special cases, like silica fume, the efficiency coefficient may actually be greater than 1. Such an approach can be applied to reconcile between specifications based on w/c ratio and concrete strength grade.

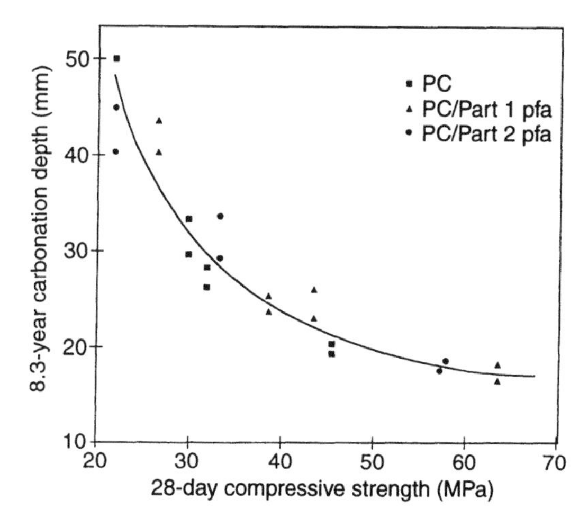


**Fig. 7.7.** Relations between the 28-day compressive strength and depth of carbonation after 8.3 years, of Portland cement concretes (PC) and concretes with 35% fly-ash (pfa) by weight of cement. (After Ref. 7.4)

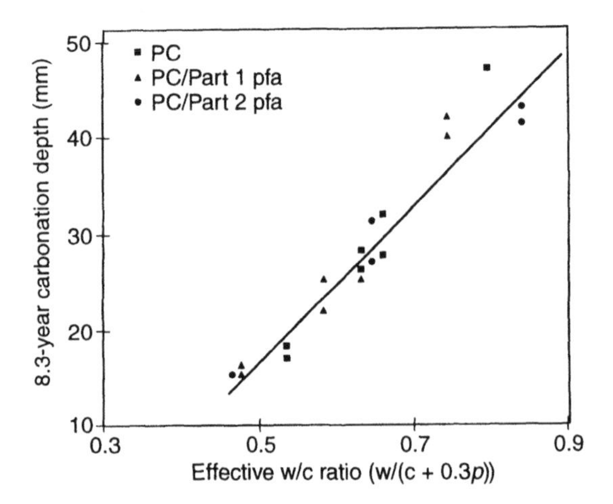


**Fig. 7.8.** Relations between effective w/c ratio (w/(c+0.3*p*)) and the depth of carbonation after 8.3 years in Portland cement concretes (PC) and concretes with 35% fly-ash (pfa) by weight of cement. (After Ref. 7.4)

### 7.2.3. Curing

Results such as those presented in the previous paragraph are based on evaluation either of concretes cured in water or concretes cured under laboratory conditions which are not harsh. The conclusions can be quite different when inadequate curing is applied, in particular in a hot–dry environment.

The combined influence of curing and w/c ratio is demonstrated in Fig. 7.9. Improper curing results in an increase in depth of carbonation by a factor of 2 to 4. It can be seen that the extent of the detrimental effects of improper curing procedures depends on the environmental conditions: they are worse under hot–dry conditions, but they can not be ignored even in the humid hot–marine environments. The need for good concrete curing practice is clearly highlighted in this figure. Prescription of concrete strength or concrete w/c ratio by itself cannot be sufficient to obtain the desired durability performance; the inadequate curing may result in ~20% reduction in the expected strength [7.8], but it will have a much greater relative detrimental influence on the depth of carbonation.

The vulnerability to curing may even be greater in concretes using blended cement where the mineral admixture is slow to react, e.g. fly-ash. This is demonstrated in Fig. 7.10 for concretes of a standard compressive strength of 33 MPa, that were moist-cured for various time periods up to 7 days. It was found that the depth of carbonation in both natural and accelerated exposures under these conditions increased with increase in fly-ash content. Similar trends were reported for the air permeability in blended cements consisting of blast furnace slag (Fig. 7.11). These trends are not to be seen as contradictory to the results of

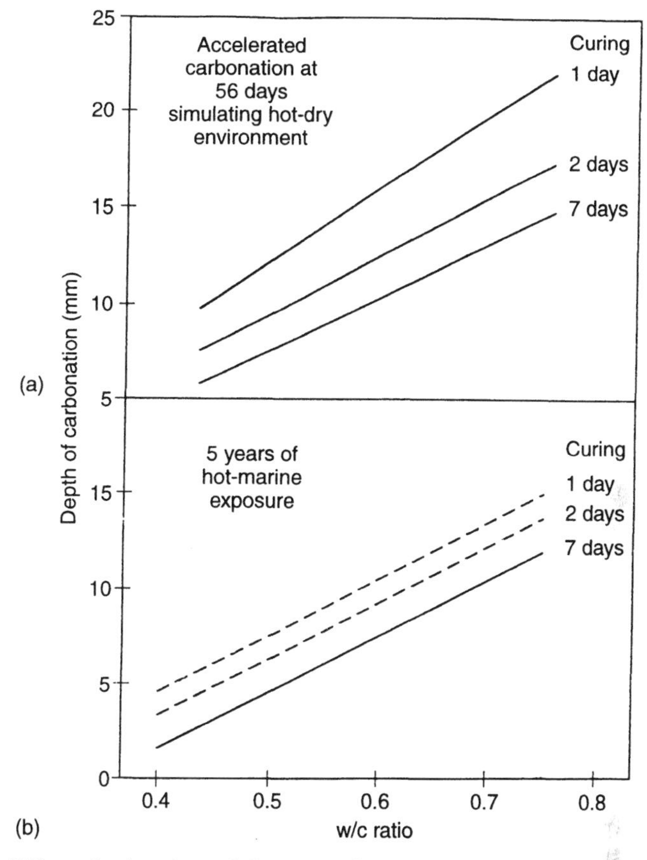


**Fig. 7.9.** Effect of w/c ratio and duration of moist curing on the depth of carbonation of concretes exposed (A) in hot–dry accelerated carbonation environment (30°C/40%RH). (Adapted from Refs 7.3 and 7.8) and (B) for 5 years to an East Mediterranean (Israel) marine environment (i.e. hot weather)

Hobbs [7.4] for properly cured concretes; rather, they highlight the influence of environmental conditions and curing procedures applied on site.

Such aspects of carbonation behaviour are indirectly referred to in some specifications which require extended curing for concretes to ensure adequate durability performance. The details of the curing needed also depend on the nature of the mineral admixture. Properties of mineral admixtures may vary considerably, even within the same family of materials called collectively by the same name (e.g. fly-ash, silica fume, slag). Such aspects of differences in behaviour are usually not addressed in specifications. Thus, if the specifications recommend the use of blended cement to improve performance, the engineer is left with the responsibility of investigating the quality of the particular mineral admixture used, and determining whether it would indeed provide the enhanced performance considering the nature of the job and the environmental conditions.

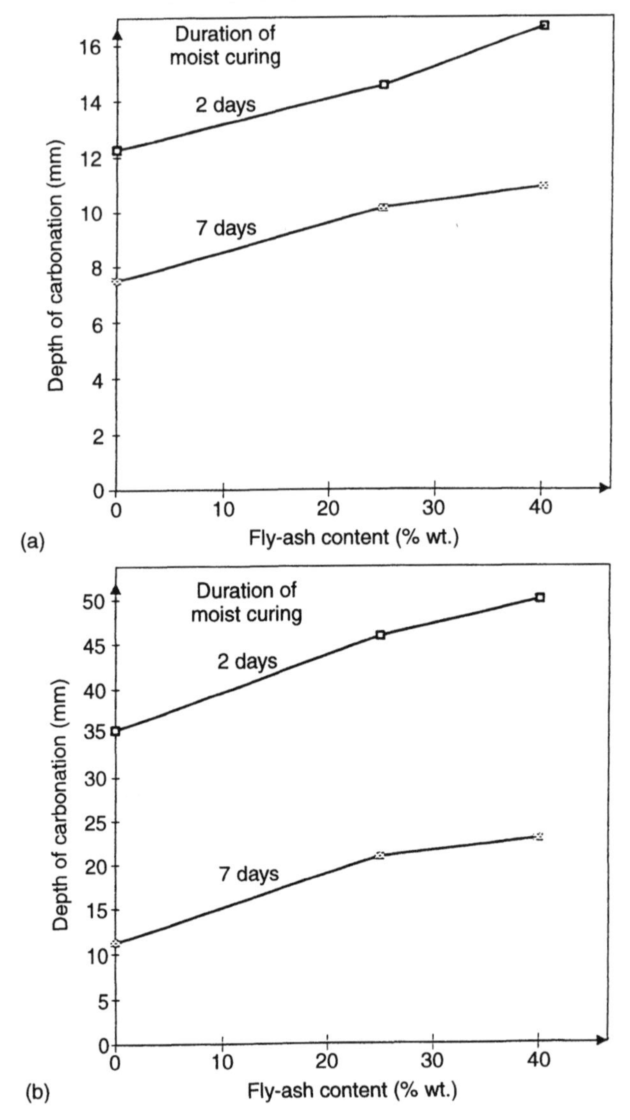


**Fig. 7.10.** Effect of moist curing on the depth of carbonation of concretes with and without fly-ash, having the same standard 28-day compressive strength (33 MPa). (A) After exposure in natural conditions for 1.5 years. (B) After accelerated carbonation for 56 days (1 day of accelerated carbonation is equivalent to about 150 days in natural exposure). (After Ref. 7.8)

Since inadequate curing has a detrimental effect on both strength and carbonation characteristics, it is often considered that strength can be used not only for specifying the quality of concrete, but also as a measure for quality control on site. Such concepts are supported by laboratory

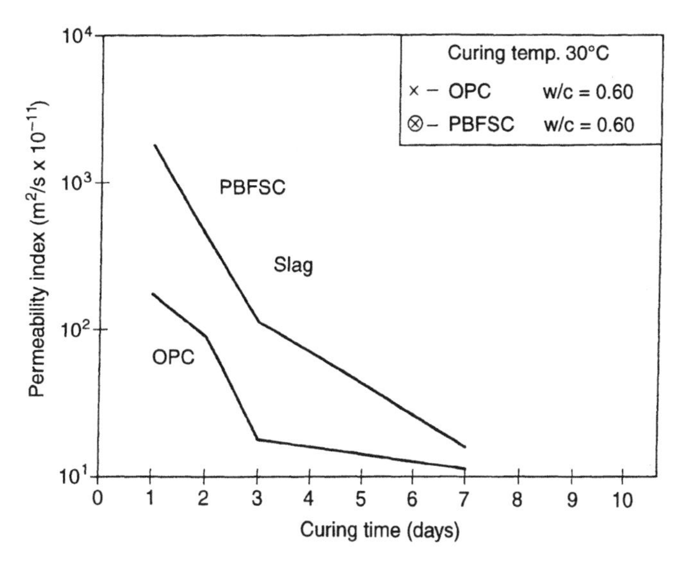


**Fig. 7.11.** Influence of blast furnace slag (68% content in the blended cement) and curing on the air permeability of 0.6 w/c ratio concretes. (After Ref. 7.9)

studies, such as that of Loo *et al.* [7.10]. They demonstrated that the rates of carbonation measured in accelerated tests (the $k$ values in eqn. (4.4) for $n = 2$) did not exhibit a universal correlation with the standard 28-day measured compressive strength (1/square root of strength – Fig. 7.12(A)), but such a correlation could be established with the actual strength of the poorly cured concrete (Fig. 7.12(B)).

Such results are in agreement with the concepts underlying those specifications that prescribe the durability performance of the concrete in terms of its strength. While this approach may be justified as a proper guideline for specifying the concrete and curing procedures (as done in reference [7.4]), it is not adequate as a quality control measure of the actual performance on site. The strength data in Fig. 7.12 (or in similar studies) were obtained for standard cube or cylinder specimens exposed to the curing environment on all sides. Deficient curing will thus affect a relatively large volume of the specimen. In the structure itself, deficient curing has an adverse effect presumably only on the exposed surface of the concrete; the core of the concrete, ~20 mm away from the surface, will remain relatively wet for prolonged periods. A core taken from a poorly cured structure may thus not exhibit the same strength retrogression as a standard size specimen exposed on all sides. This issue highlights the need to address separately the 'cover concrete' (or 'concrete skin'), where the actual moisture regime can be quite different from that which exists in the 'bulk' concrete (Fig. 7.13) [7.8, 7.11]. Several test methods have been developed to assess the properties of this part of the concrete, based on measurement of either surface absorption or of air

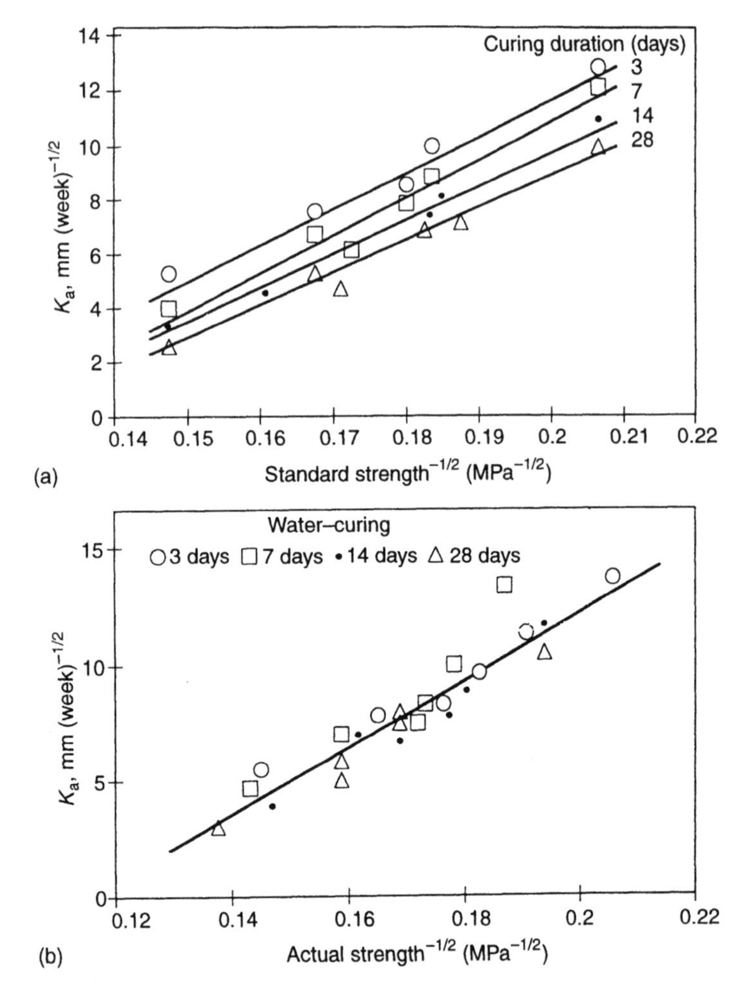


**Fig. 7.12.** Relations between strength (inverse of strength), moist curing and carbonation rate coefficient, measured in accelerated conditions. (A) Relations between standard 28-day strength and the carbonation rate coefficients. (B) Relation between the actual strength at the end of the curing period and the carbonation rate coefficient. (After Ref. 7.10)

permeability of the cover concrete [7.12, 7.13]. However, results of tests of this kind are sensitive to the test moisture condition, and currently they can not be used as a basis for specification.

## 7.3. CONTROL OF CHLORIDES

### 7.3.1. Introduction

The code requirements which refer to chloride in the concrete are intended to limit its content at the steel level. The aim is to ensure that

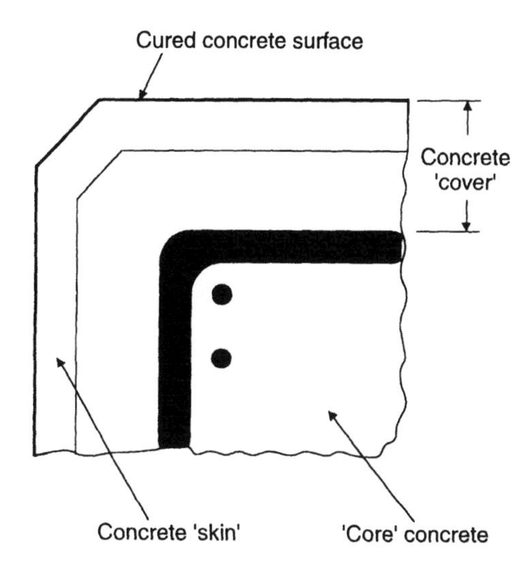


**Fig. 7.13.** Schematic description of concrete cover (concrete skin). (Adapted from Ref. 7.9)

during the service life of the concrete, the chloride content at the level of steel will remain below the critical value that induces corrosion. Alternatively, the aim is to at least delay the increase of chloride content to the critical level and ensure thereafter, in the propagation stage, that the corrosion rate will be sufficiently small to prevent damage during the service life.

It is not easy to quantify these desired concrete characteristics, as many of the factors involved in chloride corrosion can not always be defined accurately. Some of these difficulties have already been addressed in sections 4.2.2 and 4.3.1.2 in which the issues of free versus bound chlorides and the significance of chloride diffusion coefficients measured in the laboratory were discussed. In view of such limitations and the need to develop practical specifications, the approach taken is to combine data from laboratory studies and experience from field investigations in assessing the technical background needed to specify the minimum requirements for corrosion control where chlorides are the initiating agent.

### 7.3.2. Estimating Chloride Ingress

In most exposure conditions the source of chloride at the level of steel in the concrete is entry and diffusion of chloride from outside the structure. The treatment of the diffusion rate in terms of a diffusion coefficient obtained from laboratory tests has some theoretical and practical limitations [7.14]. The use of procedures based on Fick's second law (eqns (4.1) and (4.2) in section 4.3.1.1) to model the diffusion process may not be accurate, because of several reasons: (i) the non-linear chloride binding of the hydrated cement; (ii) Fick's laws are not adequate for conditions where ionic interaction between diffusing ions may take place which is the case in the concentrated pore solution in concrete; and (iii) the sensitivity of the diffusion process to the type of cation used in the test. For example, a change in the cation from sodium to calcium (i.e. diffusion measurements in NaCl and $CaCl_2$ solutions, respectively) increases the diffusion coefficient of chloride by more than a factor of 2. Another limitation is the time required to achieve steady-state conditions; when concrete specimens are tested, extremely long times are required, since concrete specimens need to be relatively large compared with the maximum aggregate size.

Rapid tests using external electrical fields to speed the migration of chloride ions (as discussed in Chapter 6) may not yield appropriate results; impressing a voltage difference produces an ionic flux which may be different from the movement of ions due to a concentration gradient, the driving force in the diffusion process.

The use of accelerated tests based on drying–wetting in chloride solutions has also been questioned as it may introduce extremely high chloride concentrations in the surface that may not be representative of natural conditions.

In view of the limitations of the various testing procedures, the current practice in developing specifications is based on synthesis of data obtained both from long-term exposure tests in natural environments and from laboratory tests. The latter are useful for comparative purposes, particularly to assess influences such as those due to the composition of the binder and w/b ratio. The former may provide better input for predicting the life span.

An approach which combines evaluation of concretes in natural exposure and the use of Fick's diffusion equations has been adopted successfully for quantitative evaluation of chloride ingress into field structures. It is based on calculation of diffusion coefficients by non-linear regression analysis. Values of $C_0$ and $D$ in eqn. (4.2) are obtained to best fit the actual chloride profile determined in the exposed concrete, taking into account the time of exposure (Fig. 7.14). The diffusion coefficient estimated by this method is usually referred to as an 'effective diffusion' coefficient, $D_{eff}$. Discussion of the conditions required to achieve adequate prediction by this method was presented in section 6.3.3.2.

### 7.3.3. Influence of Environmental Conditions

In writing specifications there is a need to take into consideration the actual environmental conditions to be experienced by the structure. These are usually specified in terms of a risk factor, which is a function of the existence of chlorides in the environment that may penetrate into the concrete. Special attention is given to splash zones in marine envi-

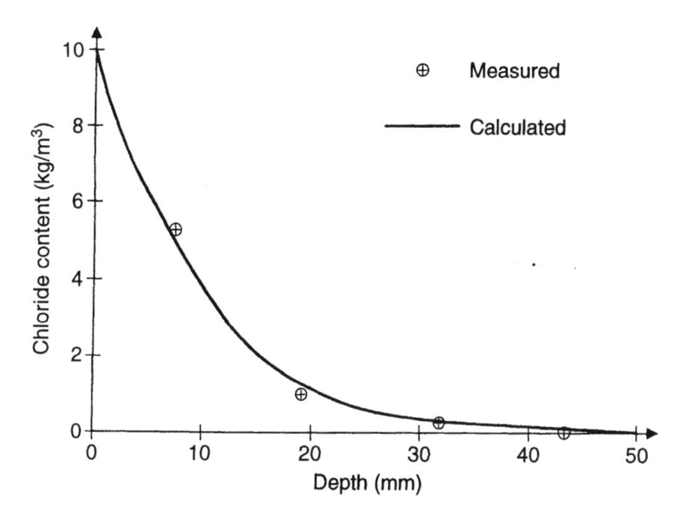


**Fig. 7.14.** Determination of the effective diffusion coefficient by fitting Fick's second law equation to the chloride content profile in concrete. (After Ref. 7.15)

ronments, and to pavements over which salt is applied to prevent freezing during winter. However, chloride may be present as aerosol form in the air away from the water front, and there is a need to define such zones, as shown in the example in Fig. 7.15. It should be emphasized that such data can not be used universally as it depends on local climatic and geographic conditions, and it may be sensitive to microclimate effects.

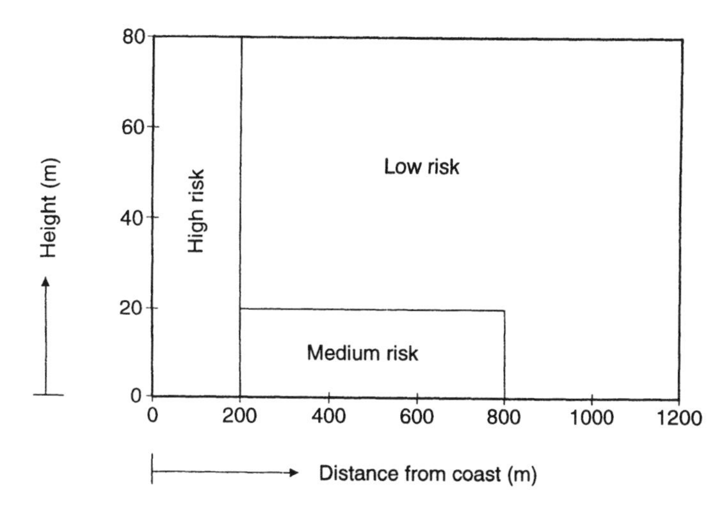


**Fig. 7.15.** Zoning of risks of due to chloride-induced corrosion as a function of the distance from the coast line and the height above ground in the Eastern Mediterranean shore (Israel), i.e. hot–marine environment. (After Ref. 7.16)

In considering risks, there is also a need to take into consideration the expected temperature range, particularly in hot–marine environments, like those existing in the Middle East. Since the diffusion process is thermally activated, the rates of diffusion are expected to increase with temperature, as predicted from eqn. (7.1):

$$D_2 = D_1 (T_1/T_2)\, e^{[k(1/T_1-1/T_2)]} \tag{7.1}$$

where $D_2$ and $D_1$ are the diffusion coefficients at temperatures $T_1$ and $T_2$, respectively.

Experimental results demonstrating the influence of temperature on the effective diffusion coefficient of high-quality concretes are presented in Table 7.1. They show an increase by about a factor of 2 for a 10°C increase in temperature.

### 7.3.4. Composition of Concrete

The combined influence of w/c ratio and composition of the cement (especially $C_3A$ content) is shown in Fig. 7.16. Rather than presenting the diffusion coefficient, the time to initiation of corrosion as estimated from electrochemical testing, is used as an indication of the durability performance. This time includes in it the sum of effects of diffusion and of binding of chlorides by the calcium aluminate hydrate product. The positive influence of increasing the $C_3A$ content of the cement is clearly evident, in agreement with the data in Fig. 4.7.

Blended cements are effective in delaying chloride induced corrosion. In well-cured systems, having the same w/b ratio, the ones with cement containing mineral admixtures have greater resistance to chloride penetration (Fig. 7.17), higher electrical resistivity [7.19, 7.20] and eventually

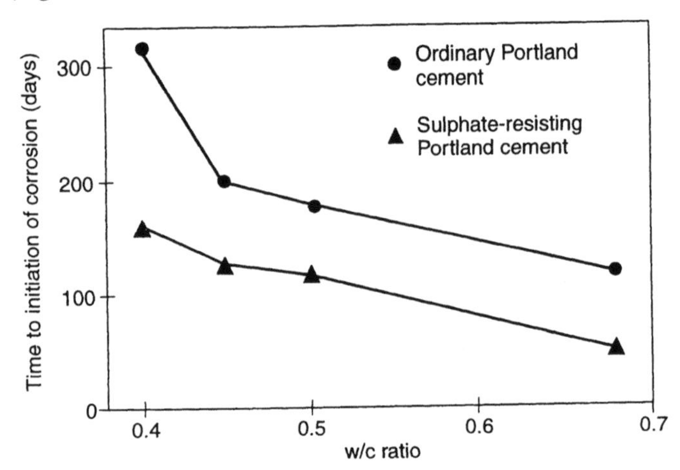


**Fig. 7.16.** Effect of $C_3A$ content and w/c ratio on the time to corrosion initiation of specimens partially immersed in 5% sodium chloride solution. (After Ref. 7.17)

**Table 7.1.** Dependence of Effective Diffusion Coefficient on Type of Concrete and Temperature.[a]

| *Type of concrete* | *Effective diffusion coefficient, ($\times 10^{-12} m^2/s$)* | |
|---|---|---|
| | *10°C* | *22°C* |
| w/c = 0.40 | 1.3 | 3.0 |
| w/c = 0.35 | 0.9 | 2.0 |
| w/b = 0.40, 7.5% silica fume | 0.6 | 1.3 |

[a] After Ref. 7.15.

lower corrosion rates (Fig. 7.18), in spite of the fact that their strength is lower [7.19].

The trend above is different from that observed for carbonation (section 7.2.2), where for the same w/b ratio carbonation-induced corrosion occurred earlier in the blended cement concrete; under these conditions the corrosion behaviour was similar when the concretes of the different binders had similar strength grade. This difference in the trends suggests that the design and specification criteria when blended cement concretes are used should not be the same for chloride- and carbonation-induced corrosion. Although many of the codes do not make this distinction, attention to such influences should be given when specifying the concrete composition.

Another important parameter in the design of concrete for durability is the cement content. The specifications usually set minimum requirements for cement content, for maximum w/c ratio, or both. The usual perception is that these two are interchangeable, as a higher cement content implies a lower w/c ratio. Although this is often the case, when it comes to corrosion control the cement content has an influence which goes beyond its effect on the w/c ratio, as seen in Fig. 7.19. The positive

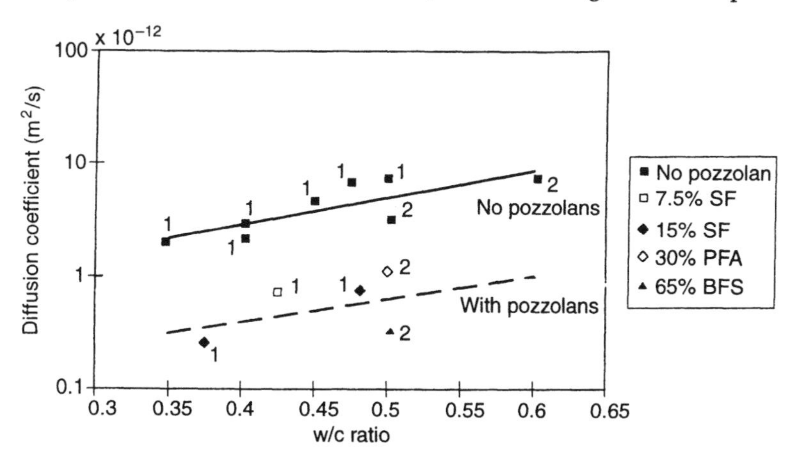


**Fig. 7.17.** Effect of pozzolans, silica fume (SF) and slag on the diffusion coefficient of chlorides in concretes of different w/c ratio. (After Ref. 7.18)

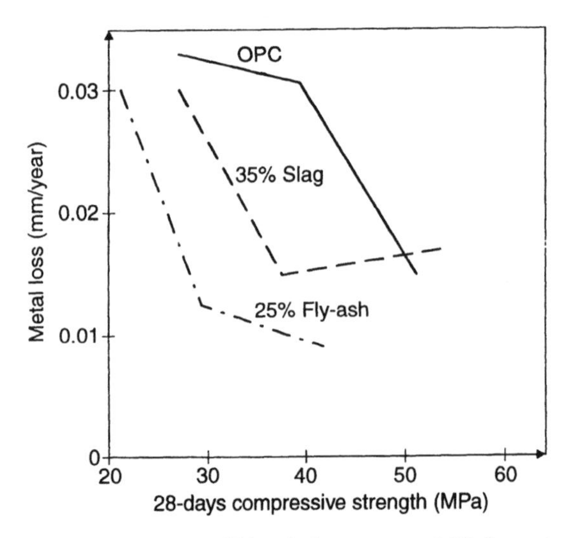


**Fig. 7.18.** Effect of composition of blended cement and 28 days strength of the concrete on the corrosion of steel in a system which was partially submerged in NaCl solution. (After Ref. 7.21)

influence of higher cement content at the same w/c ratio reflects the greater binding capacity for chloride associated with the additional cement, which slows down the ingress of chloride and leads to a lower measured effective diffusion coefficient. Note that as the w/c ratio approaches 0.4, the effect of the cement content is reduced.

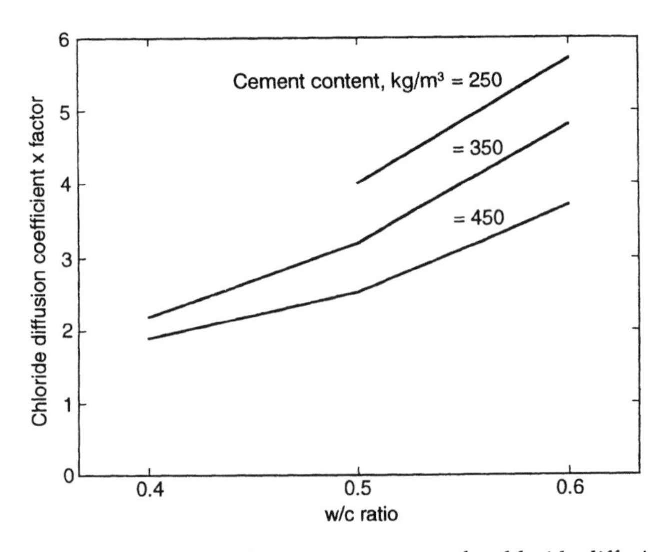


**Fig. 7.19.** Effect of w/c ratio and cement content on the chloride diffusion coefficient of concretes placed in the inter-tidal range of sea water. (After Ref. 7.22)

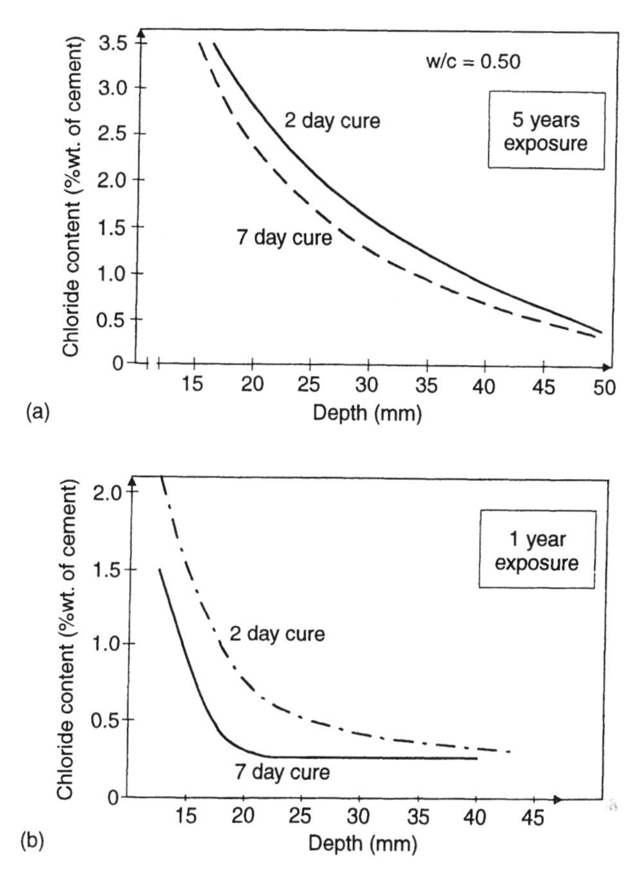


**Fig. 7.20.** Effect of curing on the chloride ingress profiles of 0.5 w/c ratio concrete after exposure for 1 and 5 years in hot–marine environment. (After Refs 7.3 and 7.23)

### 7.3.5. Curing

An important aspect of concrete technology is curing. Its influence on the chloride permeation profiles developed in concrete exposed to a hot marine environment is shown in Fig. 7.20 for 1 and 5 years of exposure of 0.5 w/c ratio concrete. It can be seen that the curing has considerable influence on the depth of chloride penetration at 1 year exposure, but at 5 years the chloride profile curves are practically the same, regardless of the initial curing. Assuming a critical chloride concentration of 0.4% of the cement weight, the depth at which corrosion can initiate was calculated, as a function of curing conditions, w/c ratio, and exposure duration (Fig. 7.21). At 1 year there is a considerable difference between the chloride profiles of the two curing regimes, but at 5 years the differences become relatively small. Inadequate curing can also have detrimental influence on resistance to chloride penetration, even in mild environments [7.24].

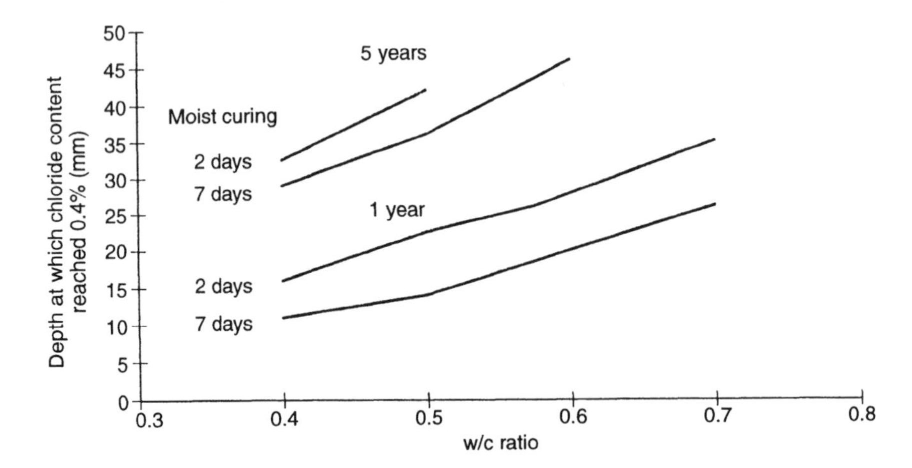


**Fig. 7.21.** Effect of curing and w/c ratio on the depth at which the chloride level reaches 0.4% by weight of cement, after 1 and 5 years exposure in hot–marine environment. (Adapted from Refs 7.3 and 7.23)

Curing effects should also be considered in relation to the influence of the binder composition in the concrete. Generally the data in the literature suggest that concretes incorporating mineral admixtures such as fly-ash, slags or silica fume are more impermeable to chloride ingress, as shown in the data in Table 4.3. However, such data is usually obtained in well-cured samples (14 to 28 days of water curing), and the trends may be different when curing is deficient. Impermeability may be reduced if curing is kept to the minimum required to achieve strength (i.e. several days of moist curing), in particular when the mineral admixture is of the type which is slow to react, like fly-ash. This can be seen in the data in Fig. 7.22, which compares the chloride penetration profiles of 0.5 w/b ratio concrete prepared with Portland cement versus concrete with 25% fly-ash blended cement, that were moist cured for 7 days only. It is clear that the fly-ash concrete is much more susceptible to chloride intrusion under these conditions.

Trends of the kind shown here suggest that, when a blended cement is recommended to improve the durability performance, attention must be paid to the nature and the quality of the particular mineral admixture, and site practices should be adjusted accordingly. As already noted in the previous section, mineral admixtures are broken down individually as fly-ash, or as natural pozzolans, or as silica fume or as metakaoline. Within each such family, the properties of specific materials may vary considerably. Since many of the specifications do not provide the means to assess the quality of the individual mineral admixture, it is part of the responsibility of the engineer to validate the quality of the chosen material. This can be done by comparative laboratory trials to assess the quality of the concrete by diffusion tests or electrochemical testing.

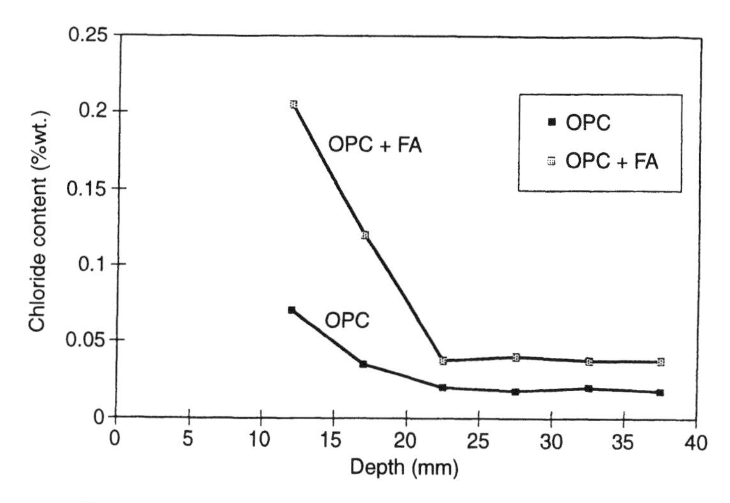


**Fig. 7.22.** Effect of short-term curing on the chloride profiles in 0.5 w/b ratio concretes with and without fly-ash (25% by weight of cement) after 1-year exposure to hot–marine environment. (Adapted from Ref. 7.23)

It is of interest to note that when the exposure conditions facilitate simultaneously chloride- and carbonation-induced corrosion, the former is likely to be the more critical. This is seen for example in the comparison between Figs 7.21 and 7.9, showing that the depth at which critical chloride level (0.4%) is developed is greater than the depth of carbonation.

## 7.4. SPECIAL PROTECTION MEASURES FOR SEVERE CHLORIDE CORROSION ENVIRONMENTS

The minimum requirements in various codes are often insufficient to ensure long-term durability of reinforced concrete in severe exposures such as found in marine splash zones, and in bridges and parking structures where de-icing salts are applied. In addition, some of the newer structures (such as commercial buildings and condominiums) built in marine areas, but not in the splash zones, are experiencing corrosion problems due to airborne chlorides. Furthermore, marine structures in the warmer climates prevalent in the Middle East, Singapore, Hong Kong, South Florida, etc. are especially vulnerable due to the high temperatures, which not only increase the rate of chloride ingress, but also the corrosion rate once the process is initiated.

In this section a brief description of supplemental corrosion protection measures is given for structures especially at risk. Good-quality concrete as described in the next subsection is considered the primary protection method, but various combinations of this with supplements are necessary to reach the desired design life of the structure.

### 7.4.1. High-Performance Concretes

One of the most effective means to increase corrosion protection is to extend the time until chloride or a carbonation front reaches the steel reinforcement. The minimal code requirements allowing the use of concrete with w/c >0.45 and concrete cover thicknesses <38 mm are totally inadequate for the structures and environmental conditions outlined above if a design life of 40 or more years is specified. In many applications, designs complying with the minimum code requirements would not provide as little as 10 years of repair-free service (see Chapter 8 for service life calculations).

Specifying much larger concrete cover thickness should enhance the service life. However, there are practical limits on cover depth; cover deeper than 50–100 mm may be impractical due to increase member sizes, loss of reinforcement effectiveness and cost for a particular structure.

A more realistic approach to be taken in such instances is the use of high-performance concrete (referred to also as high-strength concrete). High-performance concretes are essentially low w/b mixes (considerably lower than 0.45), in which the needed level of workability is achieved by using combinations of chemical admixtures (usually high-range water reducers) and mineral admixtures (usually silica fume). The use of superplasticizer in combination with cement contents of up to about 450 kg/m$^3$ and silica fume up to about 15% by weight of cement can provide workable mixes at reduced water content. In such concretes the permeability is sufficiently low as to considerably reduce the rate of chloride ingress and increase the electrical resistivity (Table 7.2) resulting in considerable reduction in corrosion (Fig. 7.23).

In long-term exposures, especially in warm environments, reduced w/c ratio and increased cover might not be adequate to meet the service life requirements. In such cases, ingress to chloride ions can be further decreased by using high-performance concretes in which, in addition to reducing the w/c ratio by use of high-range water reducing admixtures, mineral admixtures (fly-ash, silica fume or granulated ground blast furnace slag) are added.

**Table 7.2.** Effect of Water/binder Ratio (w/b), Water/cement ratio (w/c), Silica Fume Content (SF), and Compressive Strength in High-Strength Concretes on the Chloride Diffusion Coefficient and Electrical Resistance.[a]

| w/b ratio | w/c ratio | Cement content (kg/m$^3$) | SF content (% wt of cement) | 28-day compressive strength (MPa) | $D_{eff}(10^{-12}m^2/s)$ | Electrical resistance (ρ, kohmcm) |
|---|---|---|---|---|---|---|
| 0.48 | 0.48 | 347 | 0 | 35.6 | 11 | 7.7 |
| 0.41 | 0.48 | 350 | 15 | 50.7 | 0.7 | 94.7 |
| 0.38 | 0.38 | 338 | 0 | 39.9 | 2.0 | 10.8 |
| 0.31 | 0.38 | 354 | 15 | 83.6 | 0.3 | 161 |

[a]After Ref. 7.25.

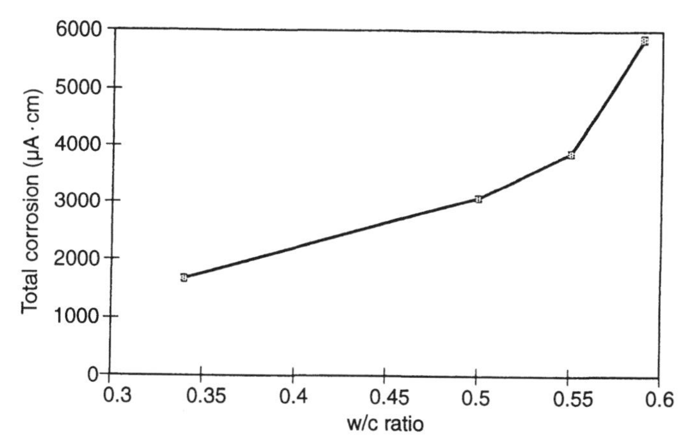


**Fig. 7.23.** Effect of w/c ratio on total corrosion of concretes ponded in chloride solution for 4 years. (Adapted from the data of Ref. 7.25)

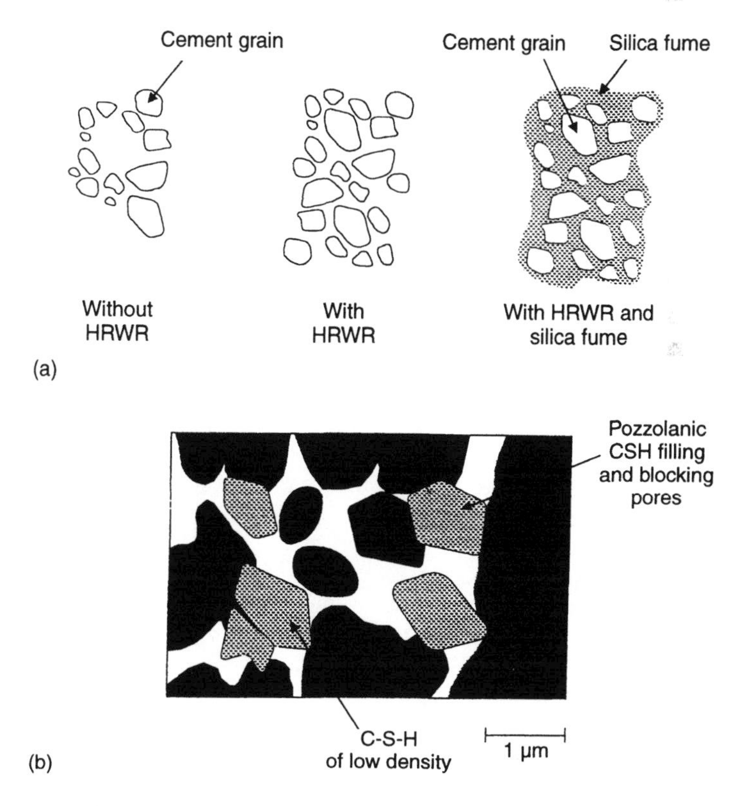


**Fig. 7.24.** Schematic description of the influence of silica fume on the structure of cementitious binder. (A) Dispersion of cement clumps by high-range water reducer and packing of silica fume particles. (After Ref. 7.26). (B) Effect of pozzolanic reaction on the microstructure of the hardened cement paste. (After Ref. 7.27)

Silica fume is the most efficient of the various types of mineral admixtures for the reduction of chloride entry into concrete. The mechanisms by which the incorporation of silica fume can produce concretes of low w/b ratio and low permeability are shown in Fig. 7.24(A). Silica fume consists primarily of spherical particles considerably smaller than ground cement particles; combining the two can provide a densely packed system. The combination of silica fume and high-range water reducer addition makes such a dense combination of particles practical. The combination of high-range water reducer and silica fume appears to function synergistically to lower the effective water demand, even beyond the reduced water demand produced by the high-range water reducer itself. Thus, a densely packed system is produced that is mobile and workable at an extremely low water content.

The chemical dispersion induced by the superplasticizer permits the individual silica fume spheres to act somewhat independently of each other; their normal state of attraction for each other ordinarily prevents such an influence. This apparently permits the silica fume particles to act somewhat as 'ball bearings' (Fig. 7.24(A)), further facilitating the development of a dense, but highly mobile, fresh concrete structure.

In addition, in the hardened concrete, silica fume acts as an efficient pozzolan, reacting quickly with the calcium hydroxide produced during cement hydration to form additional C-S-H. This response further densifies the structure of the hardened concrete, and makes the pore system more discontinuous (Fig. 7.24(B)). This results in both increased strength and reduced water and chloride permeability.

To ensure that the potential benefits of such concretes are realized in practice, it is necessary to provide prolonged and effective wet curing. This is true for silica fume concretes and slag concretes, but it is especially important for fly-ash concretes. The pozzolanic reaction of fly-ashes is slow, and it takes a long time for calcium hydroxide generated quickly in cement hydration to react with fly-ash particles. Even well-designed and well-cured fly-ash concretes tend to be somewhat more permeable than silica fume concretes, and may be less effective in delaying the onset of corrosion.

High-performance silica fume concretes possess two characteristics which impart excellent protection against corrosion of steel in concrete:

(1) Very low effective chloride diffusion coefficients, about an order of magnitude lower than those of concrete with Portland cement binder only having the same w/c ratio, and two orders of magnitude lower than Portland cement concrete of higher w/c ratio (Fig. 7.25 and Table 7.2). The result is that characteristic chloride ingress profiles, showing the combined influence of lowering w/b ratio and addition of silica fume, are like those presented in Fig. 7.26. All the concretes in this figure had the same cement content of 355 kg/m$^3$; the marked

improvements achieved with the high-performance concrete were due to the combined incorporation of high-range water reducer and silica fume.

(2) Very high electrical resistivity of the concrete, about an order of magnitude higher than that of plain Portland cement concrete (Fig. 7.27 and Table 7.2).

The overall result of these two influences is a considerable reduction in the rate and extent of corrosion of steel in concrete in chloride containing environment, as seen in Fig. 7.28.

Finally, when addressing high-performance concretes it should be emphasized that their use is not a way of substituting for good concrete practice. It is even more important to properly place, finish and cure high-performance concretes so as to realize the benefits of improved design. Procedures to prevent plastic and thermal cracking need to be in place. Proper jointing is necessary to prevent drying shrinkage cracking. Curing times should at a minimum be longer than needed to develop a discontinuous pore structure (about 3 days of water curing at 22°C for a 0.4 w/c concrete and 2 weeks for a 0.5 w/c concrete).

The *k* concept defined in section 7.2.2 for pozzolanic materials can be also applied for quantifying the influence of silica fume. The high effectiveness of silica fume shows up by *k* values larger than one, usually in the range of 2 to 3. This is different than fly-ash where the *k* value is smaller than 1, usually in the range of 0.2 to 0.4. In the discussion of suggested revisions to the European prestandard EN 206 it has been proposed to assign to silica fume which conforms to the European

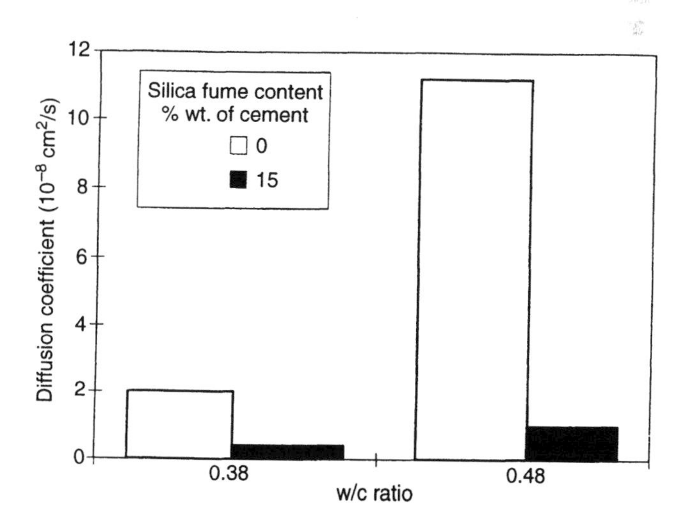


**Fig. 7.25.** Effect of silica fume and w/c ratio on the effective chloride diffusion coefficient in concretes. (Adapted from the data of Ref. 7.25)

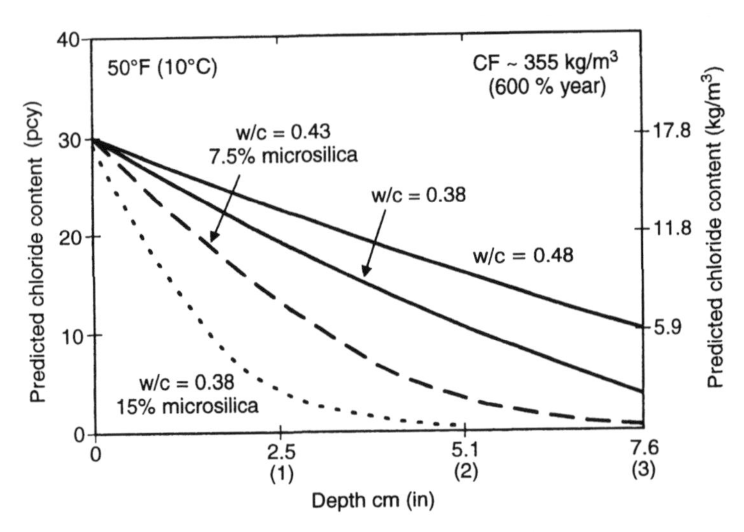


**Fig. 7.26.** Effect of w/c ratio and silica fume on the predicted chloride profiles in concretes after 50 years at 10°C. (After Ref. 7.25)

Standard EN SSS a *k* value of 2.0, except for carbonating conditions combined with wetting and drying cycles where it is reduced to 1.0.

### 7.4.2. Corrosion Inhibitors

A corrosion inhibitor for metal in concrete is a chemical substance that reduces the corrosion of the metal without reducing the concentration of

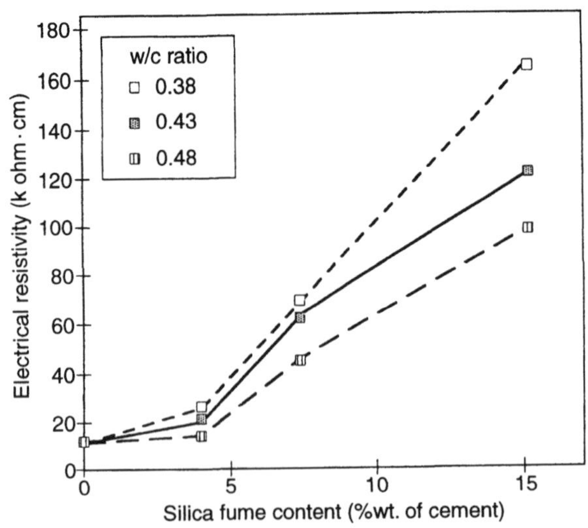


**Fig. 7.27.** Effect of w/c ratio and silica fume content on the electrical resistivity of concretes. (Adapted from the data of Ref. 7.25)

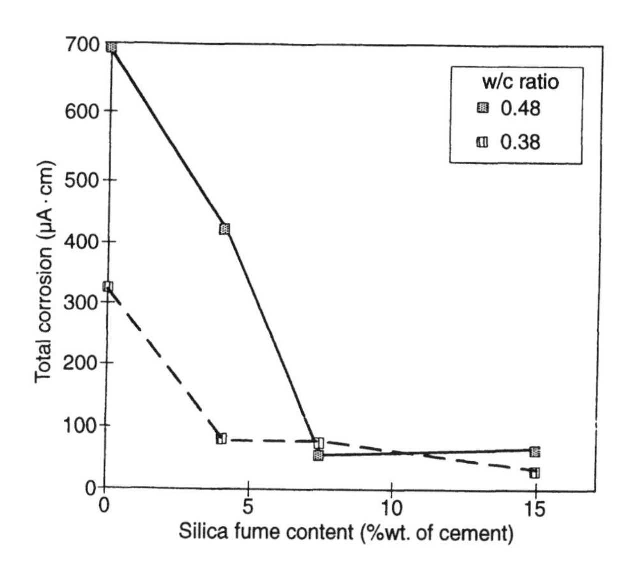


**Fig. 7.28.** Effect of w/c ratio and silica fume on the total corrosion of steel in concrete after 30 months of ponding in chloride solution. (Adapted from the data of Ref. 7.25)

the corrosive agents. This definition which is taken from ISO 8044-89 (Corrosion of Metals and Alloys-Vocabulary) makes the distinction between a corrosion inhibitor and other additions to concrete that improve corrosion resistance by reducing chloride ingress into the concrete.

Corrosion inhibitors can either influence the anodic or cathodic reactions, or even both. Since the anodic and cathodic reactions must balance each other, a significant reduction in either or both will result in a significant reduction in the corrosion rate.

Figure 7.29 illustrates the effects of both types of inhibitors when the chloride concentration has not been changed. When no inhibitors are present, the anodic ($A_1$) and cathodic curves ($C_1$) intersect at point W (Fig. 7.29(A)). Severe pitting corrosion is occurring and the corrosion rate is $I_o$. The addition of an anodic inhibitor (curve $A_2$) promotes the formation of the passivation film which raises the protection potential $E_p$, so that the anodic and cathodic curves now intersect at point X. The corresponding corrosion rate, $I_i$, is reduced by several orders of magnitude compared with $I_o$.

The addition of a cathodic inhibitor in the absence of an anodic inhibitor results in a new cathodic curve ($C_2$) (Fig. 7.29(B)). The new intersection with the anodic curve ($A_1$) is at point Y. Though the corrosion rate is reduced, pitting corrosion still occurs, because the potential remains more positive than $E_p$. Therefore, a cathodic inhibitor would have to reduce cathodic reaction rates by several orders of magnitude to be effec-

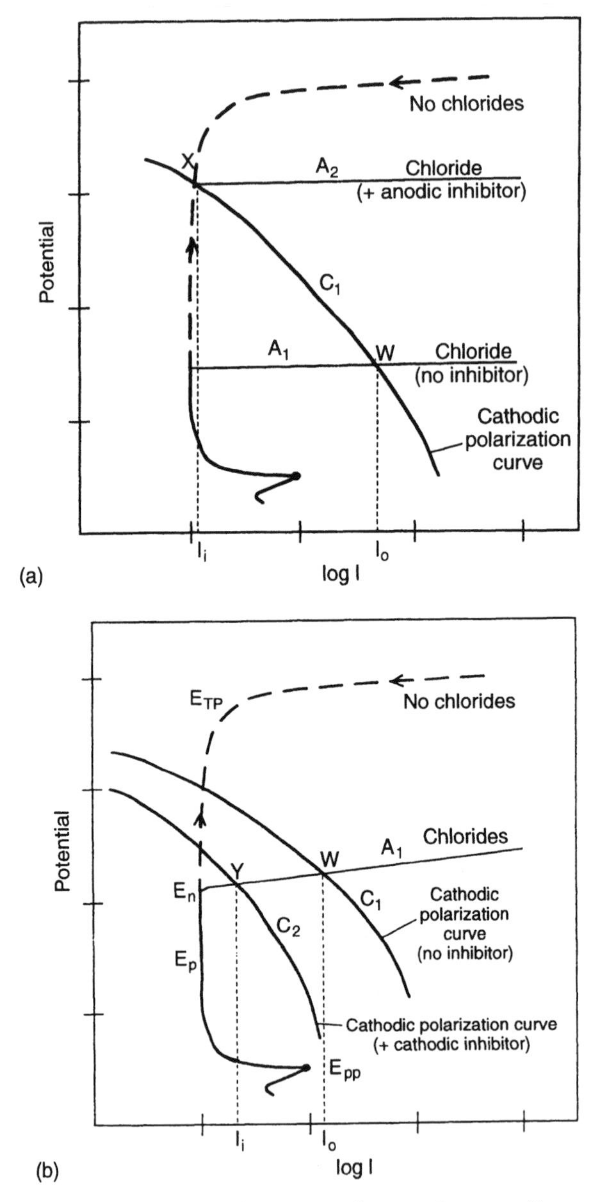


**Fig. 7.29.** Effect of inhibitors on the potential–corrosion rate diagrams of anodic and cathodic reactions. (A) Anodic inhibitor. (B) Cathodic inhibitor. $I_o$, corrosion rate without inhibitor; $I_i$, corrosion rate with inhibitor.

tive by itself. As a result, a cathodic inhibitor is usually less effective than anodic inhibitor.

Since inhibitors are added to the concrete at the mixer, it is important that they have no serious detrimental effects on the plastic or hardened properties of the concrete. For example, sodium and potassium nitrites are excellent corrosion inhibitors, but they produce severe strength regression, and the excess sodium and potassium could lead to alkali-aggregate attack for some cement and aggregate combinations. Calcium nitrite is the most widely used inhibitor and has been available since the mid 1970s.

It is important not to use a corrosion inhibitor as a substitute for good-quality concrete and practices as noted in section 7.3. The performance improvement will be minimized if high chloride contents are reached in a short time period.

Two recent reviews are available on corrosion inhibitors in concrete and are recommended for further information [7.29, 7.30].

#### *7.4.2.1. Anodic Inhibitors*

The most common anodic inhibitor is calcium nitrite. The mechanism by which it inhibits corrosion is associated with the stabilization of the passivation film which tends to be disrupted when chloride ions are present at the steel level. The destabilization of the passivation film by chlorides was discussed in section 2.3, where it was shown that it was largely due to the interference with the process of converting the ferrous oxide to the more stable ferric oxide (reactions (2.2) to (2.5) in Chapter 2). When nitrites are present they become involved in the reaction by which the ferrous ions ($Fe^{2+}$) are further oxidized to produce the more stable ferric oxide [7.31]:

$$Fe^{2+} + OH^- + NO_2^- \rightarrow NO\uparrow + \gamma\text{-}FeOOH \qquad (7.2)$$

Chloride and nitrite ions compete at 'flaws' in the passivation film (Fig. 2.4(B)) for the same ferrous ions. The relative concentrations of chloride and nitrite ions will determine the type of reaction that will take place. When the chloride ion concentration is high (i.e. large ratios of $[Cl^-]/[NO_2^-]$ and/or $[Cl^-]/[OH^-]$), the probability of $Cl^-$ complexing the ferrous ion (reaction (2.5) in Chapter 2) is increased. This forms the basis for the concept of a $[Cl^-]/[NO_2^-]$ protection ratio, beyond which pitting is likely. The ratio is in the range of 1 to 1.5 depending on the chloride content [7.31, 7.32].

Thus, nitrite is an anodic inhibitor, and it acts to raise the threshold level of chloride needed to initiate corrosion. Laboratory and field data have been used to develop a protection table shown here (Table 7.3) in which the recommended amount of calcium nitrite needed for a given level of chloride in the concrete at the level of the reinforcing steel is given. The protection claimed is based upon autopsy of actual specimens, in which it was verified that corrosion was not present at a given chloride level. An example of the benefits of calcium nitrite when good quality concrete is used is given in Fig. 7.30, demonstrating its influence in

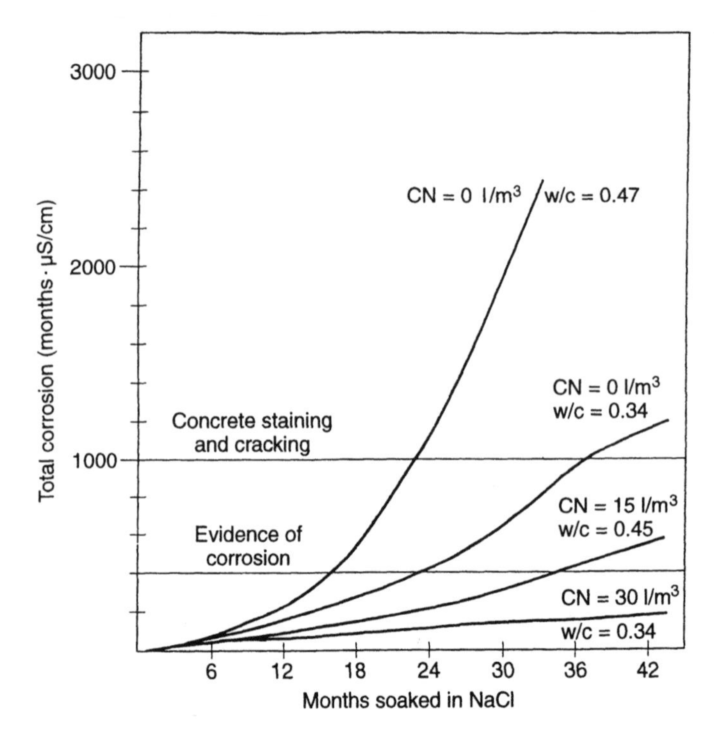


**Fig. 7.30.** Effect of w/c ratio and calcium nitrite (CN) content on corrosion of steel in concrete in NaCl-ponded specimens. (After Ref. 7.29)

systems where the w/c ratio is reduced or in systems where silica fume is added.

### *7.4.2.2. Cathodic Inhibitors*

Cathodic inhibitor adsorbs on the steel surface and thus acts as a barrier to the reduction of oxygen which is the primary cathodic reaction for steel in concrete. Screening tests on some cathodic inhibitors [7.30] using ASTM G109-92 (Test Method for Determining the Effect of Chemical

**Table 7.3.** Calcium Nitrite Dosage Required for Protection of Chloride-Induced Corrosion.[a]

| *Calcium nitrite content ($kg/m^3$, 30% sol.)* | *Chloride ion content at steel level, ($kg/m^3$)* |
|---|---|
| 10 | 3.6 |
| 15 | 5.9 |
| 20 | 7.7 |
| 25 | 8.9 |
| 30 | 9.5 |

[a] After Ref. 7.32.

Admixtures on the Corrosion of Embedded Steel Reinforcement Exposed to Chloride Environment) showed that they are able to provide modest improvement over control concrete, but not to the same extent as that obtained with anodic inhibitors.

In general, due to the passivating influence of concrete, anodic inhibitors are more efficient since they act directly to increase the level of chloride necessary to disrupt this passivity. Since the anodic corrosion is the limiting factor in the corrosion rate, only a large reduction in the cathodic reaction rate would lead to a reduction of corrosion in the presence of chloride contents sufficient to suppress pitting. As noted earlier, corrosion rates of the order of only 0.5 $\mu A/cm^2$ result in severe corrosion, and typical cathodic currents associated with oxygen reduction are in excess of 1000 $\mu A/cm^2$. Thus, a cathodic inhibitor would need to be over 99.9% efficient to prevent corrosion. Furthermore, if additional cathode is available from any source, corrosion can proceed as if the cathodic inhibitor was not present.

Thus, to achieve a sufficient level of efficiency in controlling corrosion, the content of cathodic inhibitor to be used should be high. Most cathodic inhibitors, such as amines, phosphates, zincates, and phosphonates severely increase set retardation at the levels needed to make an appreciable effect on corrosion performance.

### 7.4.3. Sealers and Membranes

Sealers and membranes have been used traditionally for providing resistance to concrete structures exposed to severe chemical attack. However, with the increase in the frequency of durability problems generated by corrosion of steel in regular reinforced concrete structures, they are increasingly used as a means for mitigating this durability problem. Membranes and sealers can provide protection by: (i) eliminating or slowing down the penetration of chlorides and carbonation, to keep the steel passivated; or (ii) reducing moisture movement into the concrete to keep it dry and slow the propagation of the corrosion reactions.

#### *7.4.3.1. Materials*

The sealers and membranes can be classified into several types [7.33, 7.34] as shown schematically in Fig. 7.31. Each type represents a family of materials with different chemical composition. Some of the more common materials within each class are presented in Table 7.4.

##### 7.4.3.1.1.Coating and sealers

These consist of continuous film applied on the concrete surface with a thickness in the range of 100 to 300 $\mu m$. The film is composed of a binder and fillers (e.g. pigments, plasticizers, catalysts, fungicides). The perfor-

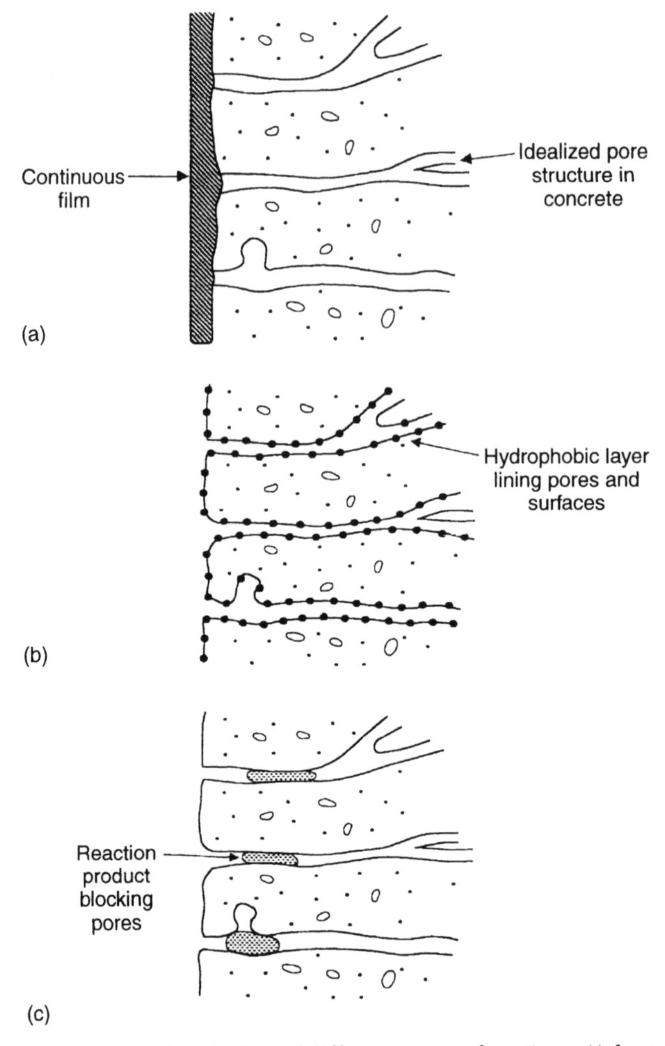


**Fig. 7.31.** Schematic description of different types of coatings. (Adapted from Ref. 7.33)

mance of the film is very sensitive to the compounding of the fillers. Therefore compositions having a similar binder may be quite different in their performance.

The coating/sealer is obtained in a liquid form which is brushed or sprayed on the concrete surface in an operation which is very much like painting. The curing process of the applied liquid, by which it converts into a film, can be based on several mechanisms: evaporation of a solvent (i.e. organic diluting liquid) or dispersant (water), chemical reaction with a catalyst that is part of the formulation, reaction with moisture in the atmosphere or in the concrete substrate, or reaction with oxygen in the atmosphere. The materials formulated for concrete coating should provide a film that is stable in the alkaline environment of the concrete.

#### 7.4.3.1.2. Pore lining

These treatments are based on processes by which the surface of the pores in the concrete are lined with materials which reduce the surface energy, to make the concrete water repellent (i.e. non-wetting). Silicone compounds are most frequently used for this purpose. Silicone resin can be dissolved in organic liquid which after evaporation deposits a film of the resin on the pore surface. Alternatively, the resin can be formulated so that it reacts with the moisture present in the pore to form the water-repellent lining. Various types of silanes have been developed for this purpose, differing in their molecular weight. Those which polymerize *in situ* to longer polymeric chains have the advantage of being less volatile and can thus offer longer service life. There is also application of water-repellent linings based on solutions of complex stearates.

#### 7.4.3.1.3. Pore blocking

These treatments are based on materials which penetrate into the pores and then react with some of the concrete constituents. The resulting products are insoluble and are being deposited in the pores to block them. The most common materials used for this purpose are liquid silicates and liquid silicofluorides. They react with $Ca(OH)_2$ to from CSH or calcium silicofluorides. Traditionally such treatments were used to strengthen concrete floors to enhance their abrasion resistance.

### *7.4.3.2. Application*

The successful performance of the coating/sealer depends not only on the quality of the materials used, but also on the application. The concrete surface should be clean and sound. Weak and cracked concrete should be removed, holes should be filled, and if necessary a levelling coat should be applied.

If the membrane applied is of a polymeric composition, the surface should be dry. Silane treatments can be done in wet concrete. However here too, it is preferred to have a dry concrete. This is needed to facilitate better penetration of the liquid compound into the pores. Coating should usually be applied in two layers to obtain a continuous film without pinholes.

The service life of the membranes/sealers ranges between 5 to 20 years [7.35]. Thus continuous maintenance and follow up is necessary, including additional treatments.

### *7.4.3.3. Carbonation Control*

The coating and sealing treatment can serve as barriers to diffusion of $CO_2$ to inhibit carbonation. The performance of the treatment can be evaluated

**Table 7.4.** Classification of Surface Treatments.[a]

| *Classification material* | | *Form of liquid paint* | *Cure* | *Colour* |
|---|---|---|---|---|
| Coating, sealer | Acrylic | Catalysed, solvented (sealer) | Loss of solvent and chemical | Clear |
| | | Solvented | Loss of solvent | Clear and pigmented |
| | | Aqueous dispersion | Drying | Pigmented |
| | | Cementitious aqueous dispersion | Drying | Pigmented |
| | Butadiene copolymer | Aqueous dispersion | Drying | Pigmented |
| | Chlorinated rubber | Solvented | Loss of solvent | Pigmented |
| | Epoxy resin | Catalysed, solvented sealer | Loss of solvent and chemical | Clear and pigmented |
| | | Catalysed, aqueous dispersion | Drying and chemical | Clear and pigmented |
| | | Catalysed, solventless | Chemical | Pigmented |
| | Oleoresinous | Solvented | Loss of solvent and oxidation | Pigmented |
| | Polyster resin | Catalysed, solvented (sealer) | Loss of solvent and chemical | Clear and pigmented |
| | Polyethylene copolymer | Solvented | Loss of solvent | Pigmented |
| | | Aqueous dispersion | Drying | Pigmented |
| | Polyurethane | Solvented (sealer) | Loss of solvent and moisture | Clear and pigmented |
| | | Catalysed, solvented (sealer) | Chemical | Clear |
| | | Catalysed | Chemical | Pigmented |
| | Vinyl | Solvented sealer | Loss of solvent | Clear |

| | | | | |
|---|---|---|---|---|
| | | Aqueous dispersion | Drying | Pigmented |
| Hydrophobic pore liner | Silicones | Solvented | Loss of solvent and reaction | Clear |
| | Siloxane | Solvented/solventless | Loss of solvent and reaction | Clear |
| | Silane | Solvented/solventless | Loss of solvent and reaction | Clear |
| Pore blocker | Silicate | Aqueous dispersion or solvented | Reaction | Clear |
| | Silicofluoride | Aqueous dispersion or solvented | Reaction | Clear |
| | Crystal growth materials | In cementitious slurry | Reaction | Cementitious |

[a] After Ref. 7.34.

by testing the resistance to carbonation of the whole system, i.e. the treated concrete, or alternatively by measuring the diffusion of $CO_2$ through a film prepared from the coating material. In the latter method a diffusion resistance coefficient, $\mu_s$, is determined. It is defined as the thickness of air layer that would provide a similar diffusion resistance as a 1-m thick layer of the coating. For a coating with a thickness of $t$, the overall resistance to diffusion is quantified in terms of the thickness of an equivalent layer of air, which would provide a similar resistance to diffusion:

$$R = \mu_s t \tag{7.3}$$

The $R$ value for $CO_2$ diffusion, $R_{CO2}$, can serve as a guidance for comparing between the effectiveness of different films. Typical values are provided in Table 7.5. Klopfer [7.37] suggested that $R$ should exceed 50 m for adequate resistance to carbonation.

It is essential that the membrane/sealer should allow diffusion of water vapour to prevent build-up of pressure that may cause blistering or delamination of the coating. Obviously, this should be achieved while maintaining efficient resistance to $CO_2$ diffusion. Klopfer [7.37] suggested that $R_{H2O}$ for the resistance to water vapour diffusion should be less than 4 m.

The factors which control the long-term performance are numerous and their evaluation is complex. They include among others the weathering performance of the coating and its bonding to the substrate. Therefore, when considering various alternatives, one should not rely only on data of the 'virgin' film, but also refer to information based on service record and experience with the coating considered, and/or data of accelerated weathering tests such as presented in Table 7.5.

**Table 7.5.** Diffusion Resistance Quantified in Terms of Equivalent Air Thickness for Several Coatings.[a]

| *Coating system* | *Initial dry film thickness (μm)* | *Equivalent air layer thickness R(m) after t hours of accelerated weathering* | | | | |
|---|---|---|---|---|---|---|
| | | *t=0* | *t=500* | *t=1000* | *t=2000* | *t=3000* |
| W-b emulsion, 80% solids heavily filled with $TiO_2$ and mica | 300 | 115 | 417 | 482 | – | – |
| As above with low viscosity methacrylate/siloxane pretreatment, unp | 300 | 150 | 210 | 127 | – | – |
| Moisture curing urethan, p.s. heavily filled | 170 | 71 | 289 | 253 | 383 | 656 |
| Methyl methacrylate, s.p. externally plasticized. 18% PVC | 200 | 42 | 150 | 332 | 283 | – |
| Urethane, two-pack, D-D s.p. | 110 | 216 | 281 | 505 | 221 | – |

| | | | | | | |
|---|---|---|---|---|---|---|
| Acrylic, w-b, translucent, low PVC | 200 | 367 | 208 | 168 | 247 | – |
| Styrene acrylate, p, 65.2% PVC, hb, s | 240 | 310 | – | 72 | 165 | – |
| Styrene acrylate, unp, w-b | 160 | 115 | 14 | 1.5 | – | – |
| Styrene acrylate, unp, w-b | 150 | 53 | 21 | 1 | – | – |
| Styrene acrylate, p, 57% PVC, s | 150 | 100 | 30 | 75 | 90 | – |
| Styrene acrylate, p, 14% PVC, s | 100 | 61 | 75 | 134 | 268 | – |
| Methylmethacrylate, p. 30% PVC, internally plasticized, s | 140 | 143 | 379 | 363 | – | – |
| Ethylene copolymer, w-b, p, 21% PVC + solvented acrylic pretreatment | 300 | 436 | 57 | 153 | 447 | – |
| Methylmethacrylate, p, 43% PVC, s | 140 | 262 | 158 | 316 | 364 | – |
| Acrylic, w-b, p | 200 | 385 | 90 | 266 | 461 | – |
| Styrene acrylate, p, s, hb | 900 | 15 | 54 | 34 | – | – |
| Alkyd, hb, s, + low viscosity sealer coat | 1000 | 57 | 10 | 19 | 45 | – |
| Acrylic, w-b, p, fibre reinforced + epoxy w-b pretreatment | 200 | 145 | 150 | 298 | – | – |
| Chlorinated rubber, hb, p.s. | 200 | 428 | 384 | 298 | – | – |

[a]After Ref. 7.36.
p, pigmented; unp, unpigmented; s, solvented; w-b, water-based; hb, high build; PVC, pigmented volume concentration.

*7.4.3.4. Resistance to Chloride Diffusion and Moisture Penetration*

Membranes and sealers can provide a barrier to penetration of moisture, to keep the concrete dry, and resistance to diffusion of chloride ions in aqueous solution, to keep the steel passivated. This kind of performance can be evaluated by a variety of tests, such as absorption tests, in which the total water absorption or rate of penetration of water is measured (i.e. surface absorption tests) and ponding tests in chloride solutions, to calculate diffusion of chlorides. Description of such tests is available in various references such as the Concrete Society Report [7.38], the American Concrete Institute Report [7.39] and references [7.40–7.42]. Data obtained in such tests is presented in Figs 7.32 and 7.33.

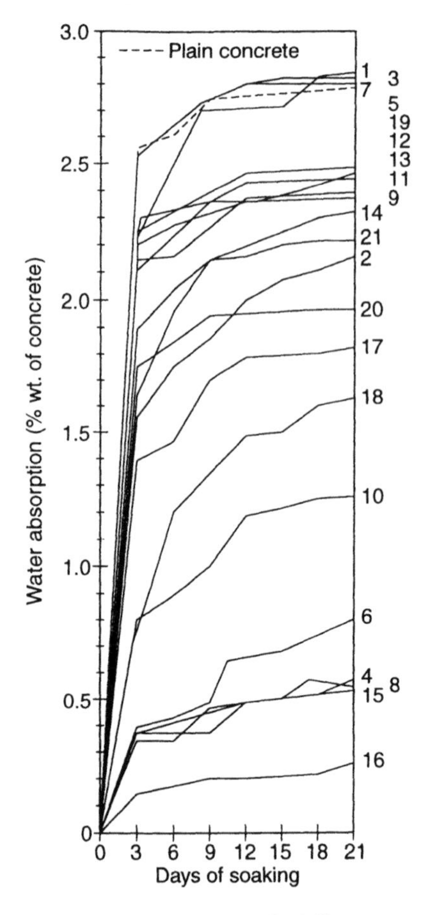


**Fig. 7.32.** Absorption curves of concretes with different coating treatments. (After Ref. 7.40). 1: siloxane; 2: boiled linseed oil; 3: siliconate; 4 and 14: two urethanes; 5: chlorinated rubber; 6: silane; 7 and 11: two materials based on butadiene; 8,10 and 13: three methacrylates; 9: silicate; 12: based on isobutylene and aluminum stearate; 15,16,18,19 and 21: five epoxies with different amounts of solids; 17 and 20: two epoxies containing polysulphides.

It should be noted that systems which performed well in tests like those presented in Figs 7.32 and 7.33 must be further evaluated for long-term performance. This could be done based on service record and experience or by evaluation in accelerated tests.

### 7.4.4. Coatings of Reinforcing Bars

One means of protecting reinforcing steel from chloride or carbonation corrosion is to coat the reinforcing bar. Sacrificial coating protects the bar by preferentially corroding and thus the reinforcing bar is protected

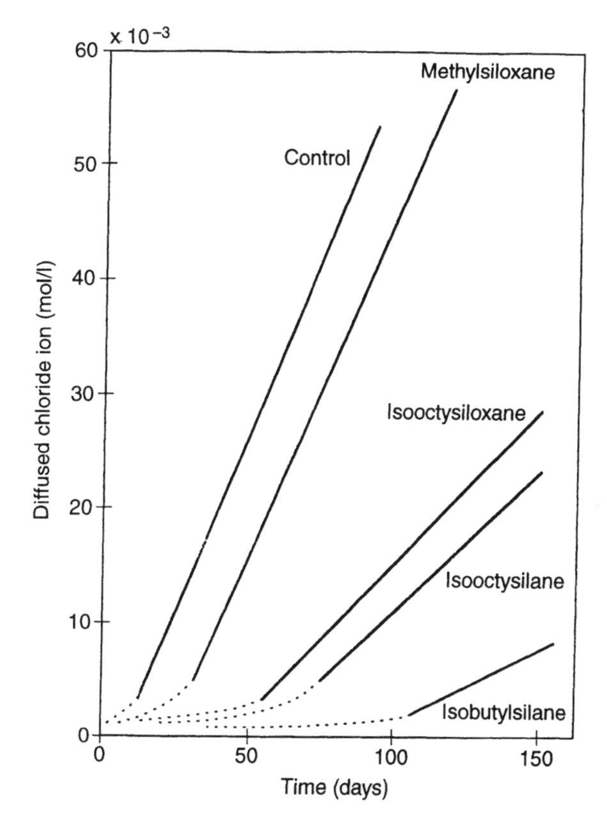


**Fig. 7.33.** Diffusion of chlorides through silane-treated concrete surfaces. (After Ref. 7.41)

galvanically. This is useful if there is a break in the coating, since the coating will still protect the exposed steel. Barrier coatings prevent the corrosive elements from reaching the steel. As long as they are maintained continuous and undamaged the steel will be protected. Barrier coatings are typically selected from materials that have low corrosion rates (noble metals such as gold, or nickel, or stainless steel) or durable non-metallic materials (polymers or ceramics). Some barrier coatings can be sacrificial to the steel and thus provide protection at breaks in the coating, and others could act as cathodes and actually increase the corrosion rate at coating flaws. The two most common coatings for steel in concrete are zinc (galvanized steel) and epoxy.

#### *7.4.4.1. Galvanized Steel Reinforcement*

Zinc coating of steel (galvanized steel) is a well-established and documented means for providing enhanced durability performance [7.43]. It acts both as a sacrificial and barrier-type coating. Of the several methods for application of zinc coating (hot-dipping, spraying, electrodeposition and diffusion), hot-dipping is the one which has been more commonly used for production of galvanized steel reinforcing bars. In this process the steel is immersed in a hot liquid zinc bath at a temperature of about 450°C. The coating deposited after cooling down consists of four layers. The external one (eta layer) is composed of pure zinc and the three below it (gamma, delta and zeta) consist of iron–zinc alloys. The thickness of each of the layers depends on the dipping conditions and the composition and treatment of the steel.

Zinc coating, like other metal coatings (and unlike polymer coating), corrodes over time. The rate of corrosion under the given environmental conditions will determine the loss of coating thickness and the time period during which it will be effective. Generally there is a fairly linear relationship between the metal thickness and the duration of its effective service life, as shown in Fig. 7.34 for galvanized steel exposed to industrial atmosphere.

When considering the effectiveness of the galvanization of reinforcing bars, it is essential to take into account the effect of the environment in the concrete to which the zinc coating is exposed, i.e. the alkaline concrete pore solution. The stability of zinc is dependent on the pH of the surrounding solution. It can be seen from Fig. 7.35 that it is stable at pH below about 12.5, but it tends to dissolve at an increasing rate as the pH increases above this level. The dissolution may be accompanied by several chemical reactions between the $Zn^{2+}$ ions and the $OH^-$ ions, leading to release of hydrogen gas and formation of zinc oxide. The oxide reacts further with $Ca^{2+}$ ions to form other complex compounds of zinc, called collectively calcium hydroxyzincate.

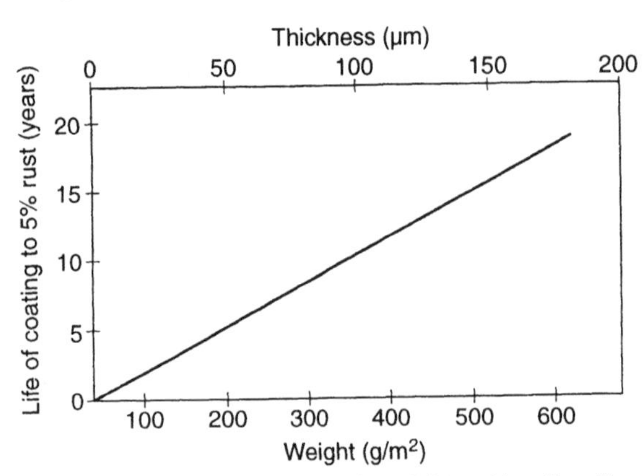


**Fig. 7.34.** Life expectancy of zinc coating in an industrial atmosphere. (Adapted from Ref. 7.43)

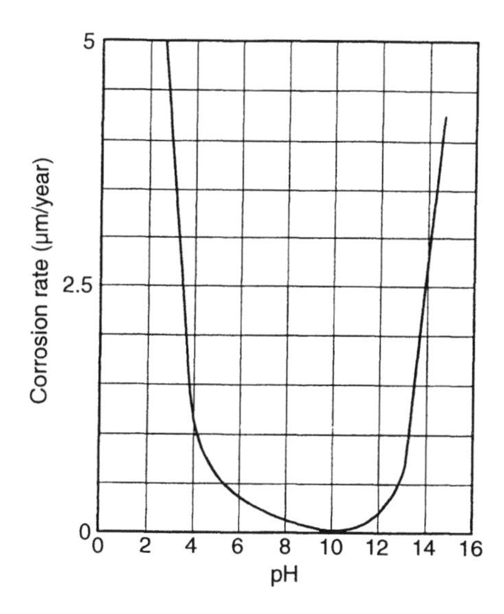


**Fig. 7.35.** Corrosion rate of pure zinc as a function of the pH of the surrounding solution. (After Ref. 7.44)

These corrosion products may be deposited at the surface of the zinc coating and seal it, thus arresting the evolution of $H_2$ gas and leading to passivation of the zinc coating. Andrade and Macias [7.45] reported that this would occur at pH values lower than about 13.3. Above this level the corrosion products are in the form of large crystals and can not effectively seal the surface. Thus, the value of pH of 13.3 can be considered as a threshold value, below which the zinc provides a passivating coating. Above a pH of 13.3 it will dissolve continuously until the coating disappears. A summary of this behaviour in the pH range of 11 to 14 is presented graphically in Fig. 7.36.

The nature of the galvanized coating and its effectiveness will thus depend to a great extent on the pH which develops in the pore solution, which in turn is a function of the alkali content of the Portland cement used. If the pH in the first few hours of hydration is below 13.3, a layer of passivating corrosion products will develop and passivate the reinforcing bar. If this layer is developed initially, a further increase in pH will not affect its stability [7.46]. However, if the pH in the initial hours is higher, a passivating layer will not form.

This sequence of events and its dependency on the alkalinity of the surrounding pore solution (shown in Fig. 7.37) may account for some of the inconsistencies reported in literature regarding the effectiveness of galvanized reinforcing bars. From Fig. 7.37 it can be calculated that, for pHs of 12.6 and 13.2, a loss of coating of 2 μm and 18 μm respectively, will occur before passivation is achieved.

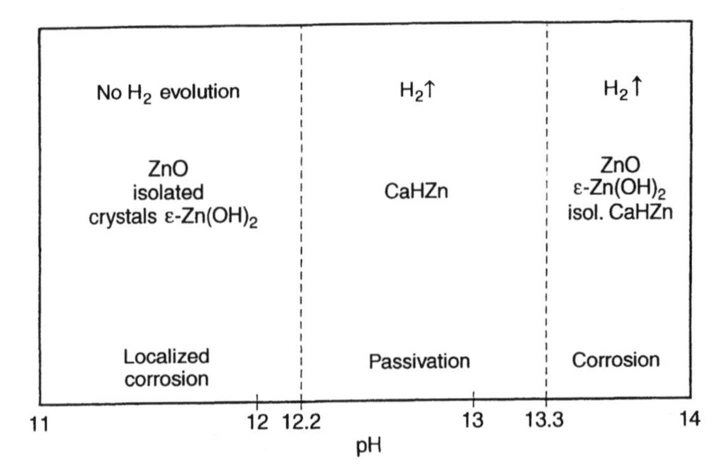


**Fig. 7.36.** Summary of the behaviour of galvanized reinforcing bars immersed in solutions with pH in the range of 11 to 14. (After Ref. 7.45)

Calculations based on tests of systems with cements of different alkali content and assumption of uniform corrosion, indicated that for a 60-μm thick coating, the service life would be 200 and 11 years for the low-alkali and high-alkali cements, respectively [7.46].

In view of such trends, the thickness of the coating should exceed 20 μm. ASTM A767/A767M-90 (Standard Specification for Zinc Coated (Galvanized) Steel Bars for Concrete Reinforcement), specifies two types of coatings, depending on their thickness: class I and class II in which the thickness is >1070 and 610 g/m², respectively (equivalent to about 85 μm and 150 μm, respectively). A limit of coating thickness of about 200 μm is

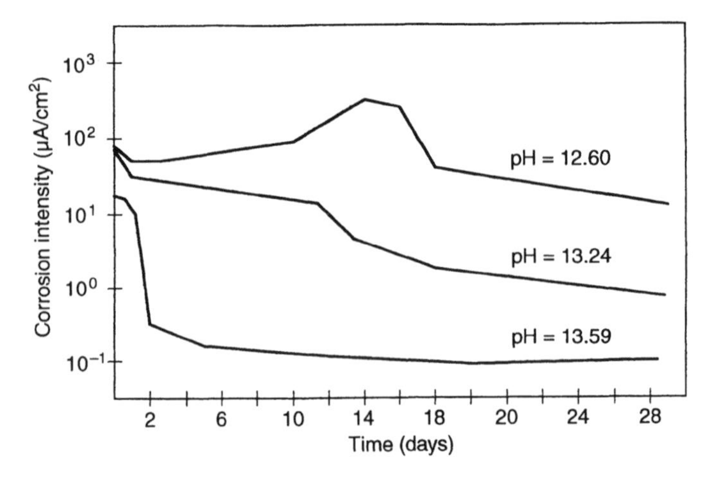


**Fig. 7.37.** Corrosion rate of galvanized reinforcing bars in solution of different pH levels. (After Ref. 7.45)

often recommended [7.47]; an increase in thickness to values which are too high may lead to reduction in the bonding between the reinforcing bars and the concrete.

In view of the dependency of the effective passivation of the zinc coating on the pH of the pore solution, the durability performance will be discussed separately for carbonation- and chloride-induced corrosion.

It was shown that at low pH values in the range 8 to 12.6 (Fig. 7.35) the zinc coating is more stable than at higher pH levels. Thus, the galvanization is expected to provide more effective protection in carbonated concrete where the pH drops to this range. This is shown in Fig. 7.38: in the un-carbonated conditions the corrosion rate is higher for the galvanized bar, but the rate might be considered practically negligible for both, being less than about 0.1 μA/cm². After carbonation, the corrosion rate increased considerably in the unprotected steel, to a level of about 2 μA/cm², whereas the increase in the galvanized bar was small, to a value of about 0.2 μA/cm². Therefore galvanization is expected to be an effective protective means in carbonation-induced corrosion.

There is controversy regarding the effectiveness of galvanization in chloride-induced corrosion. The conflicting views stem from the fact that galvanization can not prevent chloride corrosion and pitting, but only delay it and slow the rate of corrosion. This is demonstrated in Fig. 7.39, which shows that the corrosion rate of the galvanized steel is substantial, although lower than that of the bare, 'black' steel. Yeomans [7.48] suggested that the galvanization can increase the threshold chloride content by 150–250%, and extend the time of corrosion initiation by a factor as high as 4.

It should be noted that when galvanized and bare steel reinforcing bars are used together, galvanic corrosion may be generated between the

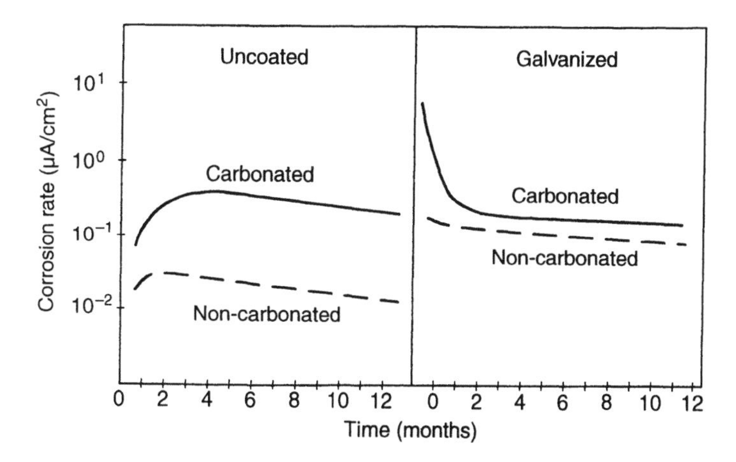


**Fig. 7.38.** Corrosion rates of galvanized and bare ('black') reinforcing bars in carbonating and non-carbonating conditions. (Adapted from Ref. 7.46)

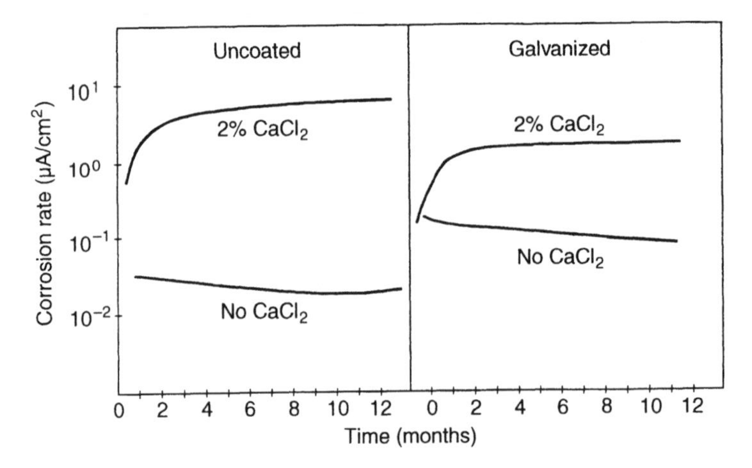


**Fig. 7.39.** Corrosion rates of galvanized and bare ('black') reinforcing bars in chloride and chloride-free environments. (Adapted from Ref. 7.46)

two, thus accelerating the corrosion processes. Attention should be given to such influences; if galvanized bars are used, all the reinforcement should be of this kind. Some of the conflicting data on performance of galvanized bars may be due to such influences; early studies sponsored by the USA Federal Highway Administration (FHWA) in which galvanized bars were coupled with bare bars, indicated poor performance [7.49]. However, later studies using galvanized bars only, that were not coupled, showed the advantages of galvanization [7.50].

In view of the advantages and limitations of galvanization, it is perhaps worthwhile to follow some of the recommendations of the Concrete Institute of Australia [7.51]:

- Galvanization should be used to provide increased protection rather than as a substitute for proper concrete quality and cover.
- Ensure that the passivating solution is of the correct strength.
- Avoid mixing galvanized and ungalvanized reinforcement.
- Check for damage to the coating after bending and increase the bend diameters if possible.
- Take extra precautions when welding galvanized reinforcement.

#### *7.4.4.2. Epoxy-Coated Steel*

Epoxy coating of steel can enhance the durability performance by serving as a barrier preventing the access of aggressive species to the steel surface and providing electrical insulation. Epoxy coating can be applied in various ways, either as a liquid or as a powder which is fused on the surface. The permeability through the epoxy coating is smaller when it is applied as a fused powder (Fig. 7.40) Therefore this method is the more common one used for production of epoxy-coated reinforcing bars. The coating is applied on the straight bar, and therefore it should be sufficiently flexible to prevent cracking when the bar is bent. The thickness of the coating is in the range of 130 μm to 300 μm, which provides optimum performance: it is sufficiently thin to allow flexibility in bending and is thick enough to be an effective barrier [7.46]. The effectiveness of the coating would also depend on its continuity and lack of defects. Therefore, special attention should be given to the coating process to achieve these characteristics.

ASTM A775M/775M-93 (Specification for Epoxy-Coated Reinforcing Steel Bars), addresses these issues by presenting the following requirements:

- The coating thickness should be in the range of 130 to 300 μm.
- Bending of the coated bar around a standard mandrel should not lead to formation of cracks.
- The number of pinhole defects ('holidays') should not be more than six per metre.
- The damaged area on the bar should not exceed 2%.

Since defects in the coated bars may be generated in the handling during transportation to the site and in the site itself during the storage and construction stages, several guidelines for proper workmanship have been established and are well documented [7.46, 7.53, 7.54]. They address good practices for delivery, unloading, storage, field cutting, bending and straightening, and placing and consolidation of the concrete. The latter

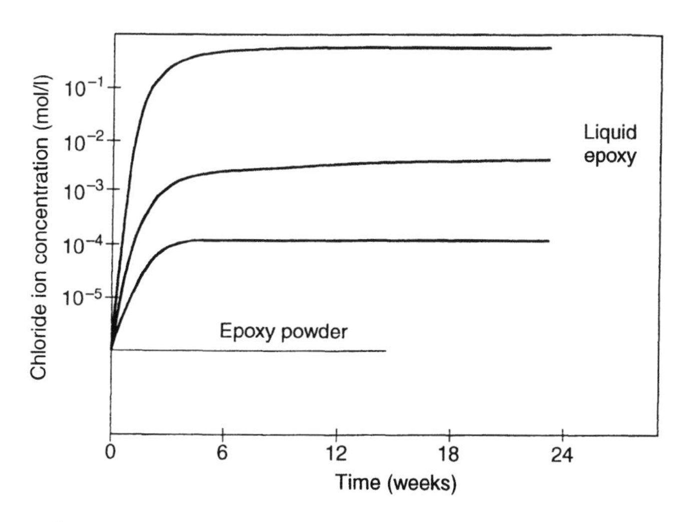


**Fig. 7.40.** The concentration of chloride ions which penetrated through epoxy coating film of 76 μm thickness. (After Ref. 7.52)

stage is particularly troublesome, as the damage that may be caused can not be inspected or repaired.

The epoxy coating reduces the bond between the reinforcing bar and the concrete. Therefore design procedures have been developed to account for this influence. These include increase in the lap length by 20–50% [7.55].

The application of epoxy-coated reinforcing bars was the result of studies sponsored by the Federal Highway Administration in the USA (FHWA) which evaluated the performance of this coating in chloride environments. These conditions exist in bridge decks where chloride de-icing salts are applied during winter. Numerous studies indicated the benefits of this coating in chloride environments (e.g. references [7.56–7.58]) as demonstrated by data such as shown in Fig. 7.41. Based on such data epoxy-coated reinforcing bars have been specified for bridge decks and also for marine structures since the early 1970s.

However, in recent years failures have been reported, most notably in bridge columns exposed to sub-tropical marine environment in the Florida Keys [7.59, 7.60]. Inspection of these structures suggested that the epoxy coating, although intact, disbonded from the reinforcing bar, even in zones where corrosion did not occur [7.60]. At present, the cause of this problem is not clear, and investigations are being carried out to determine whether it is an inherent limitation of the epoxy coating itself

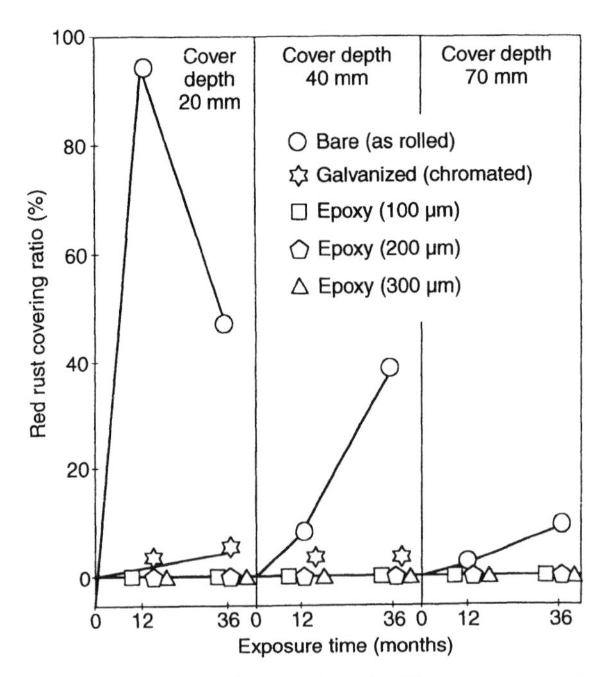


**Fig. 7.41.** Rusting of reinforcing bars with and without epoxy coating for different depth of concrete cover. (After Ref. 7.63)

(e.g. permeation of chlorides even in undamaged coating), or the result of poor practice which caused defects, either in the production of the coating in the plant, or during the handling (e.g. exposure to aggressive environment during storage) and the construction stages.

There has always been a concern with holidays and flaws in the coating which can be sites for corrosion initiation, and with disbondment of the coating from the steel over time. These problems have recently been discussed in TRB and ASTM symposium [7.61]. The development of primers and production methods that will improve the epoxy–steel bond were highlighted in these meetings. However, it was also pointed out that placing and vibration of concrete over the epoxy-coated bars could also result in flaws.

In view of these uncertainties, it is perhaps adequate to adopt the strategy suggested for galvanized reinforcement, namely that epoxy coating should be used to provide additional protection rather than as a substitute for proper concrete cover and quality.

### 7.4.5. Cathodic Protection

Cathodic protection, if applied properly, can prevent corrosion of steel in concrete, and stop corrosion that is already in progress. It accomplishes

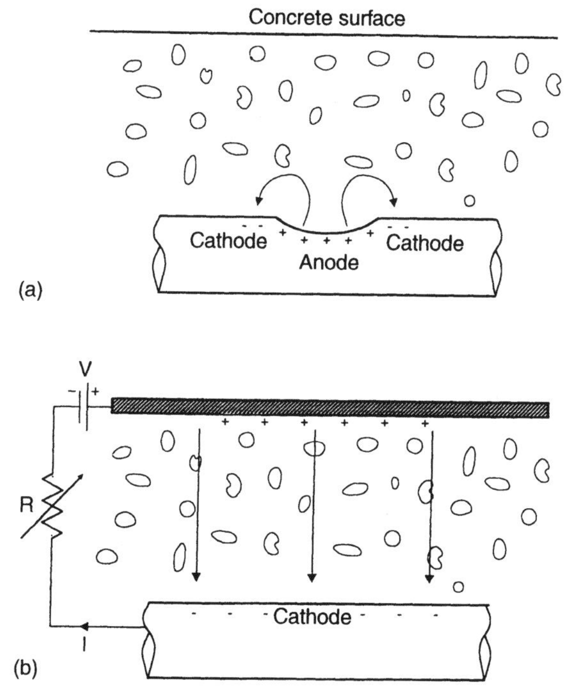


**Fig. 7.42.** Schematic description of cathodic protection.

this by making the steel reinforcing bar a cathode by use of an external anode, as shown schematically in Fig. 7.42. Electrons are supplied to the reinforcing bar from the anode, through the ionically conductive concrete. The current supplied should be sufficiently high so that all the local cells are inhibited and all the steel surface becomes anodic.

The external current can be supplied by connecting the steel to a metal which is higher in the electrochemical series (e.g. zinc). It serves as the anode relative to the cathodic steel. In this method the anode gradually dissolves as it oxidizes and supplies electrons to the cathodic steel. Therefore this type of cathodic protection is referred to as 'sacrificial protection'. The anode is called a 'sacrificial anode'.

An alternative method for cathodic protection is based on supplying electrons to the reinforcing steel from an external electrical power source. The electrical power is fed into an inert material which serves as the anode and is place on the concrete surface. This method is referred to as 'impressed current anodic protection'. The anode is frequently called a 'fixed anode'.

A sacrificial anode can be particularly effective in submerged structures where the concrete is wet and the resistivity is low; the relatively small potential difference of the two metals is sufficient to render cathodic protection. In structures exposed to the atmosphere, the electrical resistance is usually higher and therefore a greater potential difference is required. This can be more readily achieved by impressed current anodic protection.

A schematic description of the layout of an impressed current cathodic protection circuit is provided in Fig. 7.43. The fixed anode can be in the form of a corrosion-resistant wire mesh placed over the concrete surface and sprayed with a conductive paint. The anode, the steel reinforcing bars, the power unit and the reference electrode are all connected by cables. The cables and the fixed anode are then protected with a cementitious coat. Measurements of the potential difference between the reference electrode and the steel can provide information as to the effectiveness of the protection, and can serve as a guide to adjust the power input. The power unit is essentially like a small commercial battery charger.

A recent RILEM report [7.62] has emphasized several considerations to assure efficient cathodic protection:

- The reinforcement must be electrically continuous; discontinuities should be corrected by providing additional connections.
- The concrete between the steel and the anode should provide electrolytic conductivity.
- Alkali-reactive aggregate should be avoided, as the cathodic protection may aggravate the alkali–aggregate reactivity problem.

Attention should be given to certain problems that may arise due to cathodic protection. The ionic current in the concrete pore solution

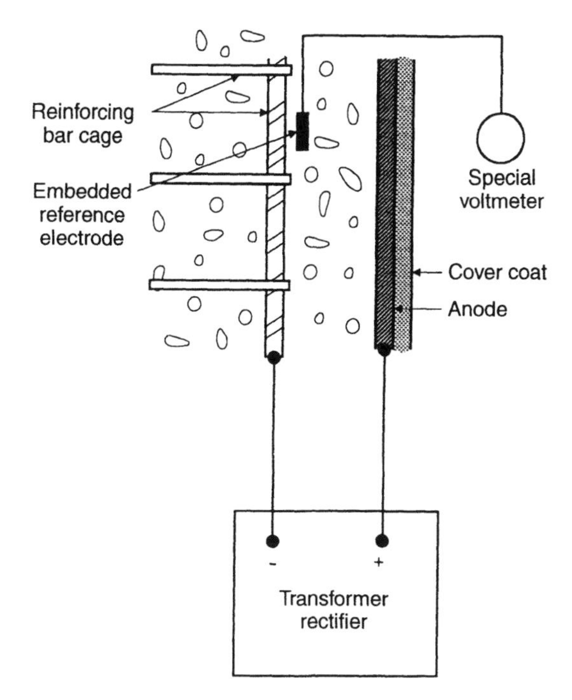


**Fig. 7.43.** The electrical circuit in impressed current cathodic protection (after [7.63]).

causes the migration of anions (e.g. $Cl^-$, $OH^-$) towards the anode away from the steel, and the corresponding movement of cations (e.g. $H^+$) towards the steel (Fig. 7.44). Removal of $Cl^-$ from the steel is an obvious advantage, and forms the basis for the development of specific electrochemical methods for chloride removal. However, the reduced alkalinity around the steel (i.e. removal of $OH^-$ ions) may negatively affect the bond between the steel and the surrounding concrete and cause delamination. Also, the increase in $H^+$ ion concentration may lead to hydrogen embrittlement, in particular in prestressed wires. It has therefore been recommended not to use this method of protection in prestressed concrete [7.63]. In view of these problems there is a need to monitor cathodic protection systems continuously, to assure that they provide effective protection without aggravating side effects. In a recent review of field cathodic protection it has been noted that many cases of poor performance were the result of improper procedures [7.64].

Cathodic protection is an expensive method. In new construction it is likely that it would be less cost effective than improving the concrete quality. However, cathodic protection has significant advantages for repair purposes, in particular in chloride-induced corrosion, where it is difficult to restore the passivation characteristics of the concrete cover.

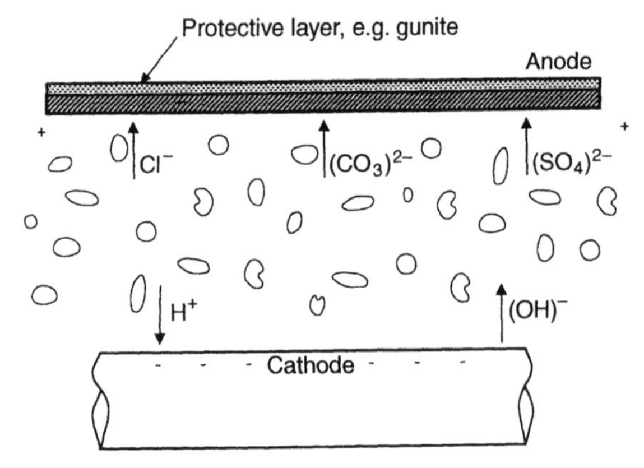


**Fig. 7.44.** Migration of ions in cathodically protected reinforced concrete (after [7.63]).

## REFERENCES


7.1. Wierig, H.J., Longtime studies on the carbonation of concrete under normal outdoor exposure. In *Proc. RILEM Seminar on Durability of Concrete Structures Under Normal Exposure*, Universitat Hannover, Hannover, 1984, pp. 239–53.

7.2. Parrott, L.J., Moisture profiles in drying concrete. *Advances in Cement Research*, **1**(3) (1988), 164–70.

7.3. Jaegermann, C. & Carmel, D., Factors affecting the penetration of chlorides and depth of carbonation. Research Report 1984–1987, Building Research Station, Technion-Israel Institute of Technology, Haifa, Israel, 1988 (in Hebrew with English Summary).

7.4. Hobbs, D.W. Carbonation of concrete containing PFA. *Mag. Conc. Res.*, **46**(166) (1994), 35–8.

7.5. Cao, H.T. & Bucea, L., Corrosion behaviour of steel embedded in fly-ash blended cements. In *Durability of Concrete*, ed. V.M. Malhotra. ACI SP-145, American Concrete Institute, Detroit, 1994, pp. 215–27.

7.6. Horiguchi,K., Chosokabe, T., Ikabata, T. & Suzuki, Y., The rate of carbonation in concrete made with blended cement. In *Durability of Concrete*, ed. V.M. Malhotra. ACI SP-145, American Concrete Institute, Detroit, 1994, pp. 917–31.

7.7. Barker, A.P. & Mathews, J.D., Concrete durability specification by water/cement or compressive strength for European cement types. In *Durability of Concrete*, ed. V.M. Malhotra. ACI SP-145, American Concrete Institute, Detroit, 1994, pp. 1135–59.

7.8a. Bentur, A. & Jaegermann, C., Effect of curing and composition on the development of properties in the outer skin of concrete. *ASCE J. Materials in Civil Engineering*, **3**(4) (1990) 252–62.

7.8b. Bentur, A. & Jaegermann, C., Effect of curing in hot environment on the properties of the concrete skin. Research Report, Building Research Station, Technion-Israel Institute of Technology, Haifa, Israel, 1989 (in Hebrew).

7.9. Schonlin, K. & Hilsdorf, H., Evaluation of the effectiveness of curing of concrete structures. In *Concrete Durability*, ed. J.M. Scanlon. ACI SP-100, American Ceramic Society, Detroit, 1987, pp. 207–26.

7.10. Loo, Y.H., Chin, M.S., Tam, C.T. & Ong, K.C.G., A carbonation prediction model for accelerated carbonation testing of concrete. *Mag. Conc. Res.*, **46**(168) (1994), 191–200.

7.11. Meyer, A. The importance of surface layer for the durability of concrete structures. In *Concrete Durability*, ed. J.M. Scanlon. ACI SP-100, American Ceramic Society, Detroit, 1987, pp. 49–61.
7.12. Dhir, R.K., Jones, M.R., Byars, E.A. & Shaaban, I.G., Predicting concrete durability from its absorption. In *Durability of Concrete*, ed. V.M. Malhotra. ACI SP-145, American Concrete Institute, Detroit, 1994, pp. 1177–94.
7.13. Kropp, J., Performance criteria for concrete durability. State of the art report prepared by RILEM technical committee TC 116-PCD, E & FN SPON, UK, 1995.
7.14. Gjorv, O.E., Performance criteria for concrete durability. In *Proc. Int. Workshop on Rational Design of Concrete Structures Under Severe Conditions*, Hokadate, Japan, 1995, pp. 51–79.
7.15a. Berke, N.S. & Hicks, M.C., Estimating the life cycle of reinforced concrete decks and marine piles using laboratory diffusion and corrosion data. In Corrosion Forms and Control for Infrastructure, ed. V. Chaker. ASTM STP 1137, American Society of Testing and Materials, Philadelphia, 1992, pp. 207–31.
7.15b. Berke, N.S. and Hicks, M.C., Predicting chloride profiles in concrete. In *Corrosion-93, the NACE Annual Conference*, National Association of Corrosion Engineers, Houston, Texas, 1993, paper 341.
7.16. Jaegermann, C., Baum, H. & Carmel, D., Influence of the distance from the sea shore on the chlorides penetrating into concrete. Research Report, National Building Research Institute, Technion-Israel Institute of Technology, Haifa, Israel, 1992 (in Hebrew).
7.17. Rasheeduzzafar, F.D., Al-Saaadoum, S.S., Al-Gathani, A.S. & Dakhil,F.H., Effect of tricalcium aluminate content on corrosion of reinforcing steel in concrete. *Cement Concrete Res.*, **20**(5) (1990), 723-38.
7.18. Gjorv, O. & Vennesland, O., Evaluation and control of steel corrosion in offshore structures. In *Concrete Durability*, ed. J.M. Scanlon. ACI SP-100, American Ceramic Society, Detroit, 1987, pp. 1575–602.
7.19 Roper, H., Sirivantnanon, V. & Baweja, D., Long-term performance of Portland and blended cement concretes under marine conditions. In *Durability of Concrete*, ed. V.M. Malhotra. ACI SP-145, American Concrete Institute, Detroit, 1994, pp. 331–51.
7.20. Rasheeduzzafar, F.D. & Mukarram, K., Influence of cement composition and content on the corrosion behaviour of reinforcing steel in concrete. In *Concrete Durability*, ed. J.M. Scanlon. ACI SP-100, American Ceramic Society, Detroit, 1987, pp. 1477–502.
7.21. Guirguis, S., Cao, H.T. & Baweja, D. Minimizing corrosion of steel reinforcement-implementation of research into practice, In *Durability of Concrete*, ed. V.M. Malhotra. ACI SP-145, American Concrete Institute, Detroit, 1994, pp. 263–81.
7.22. Pollock, D.J., Concrete durability tests using the Gulf environment. In *Proc. 1st Int. Conf. On Deterioration and repair of reinforced concrete in the Arabian Gulf, Bahrain, 1985*, The Bahrain Society of Engineers, Bahrain, Vol. I, pp. 427–41.
7.23. Jaegermann, C. and Carmel, D., Factors affecting the penetration of chlorides and depth of carbonation. Final Research Report, National Building Research Institute, Technion-Israel Institute of Technology, Haifa, Israel, 1991 (in Hebrew).
7.24. Senbetta, E. and Malchow, G., Studies on control of durability of concrete through proper curing. In *Concrete Durability*, ed. J.M. Scanlon. ACI SP-100, American Ceramic Society, Detroit, 1987, pp. 73–88.
7.25a. Berke, N.S. & Sundberg, K.M., The effects of admixtures and concrete mix designs on long-term concrete durability in chloride environments. In

*Corrosion-89*, The NACE Annual Conference, National Association of Corrosion Engineers, Houston, Texas, 1989, paper 386.

7.25b. Berke, N.S. and Sundberg, K.M., The effects of calcium nitrite and microslica admixtures on corrosion resistance of steel in concrete. In *Performance of Concrete*, ed. D. Whiting. ACI SP-122, American Concrete Institute, Detroit, 1990, pp. 265–280.

7.26. Bache, H.H., Densified cement ultrafine particle based materials. Paper presented at 2nd International Conference on Superplasticizers in Concrete, Ottawa, 1989.

7.27. Mehta, P.K. & Monteiro, P.J.M. *Concrete: Structure Properties and Materials*. Prentice-Hall, 1993.

7.28 Berke, N.S. & Sundberg, K.M., The effect of calcium nitrite and microsilica admixtures on corrosion resistance of steel in concrete. in *Performance of Concrete*, ed. D. Whiting. ACI SP122, American Concrete Institute, 1990, pp. 265–80.

7.29. Berke, N.S., Corrosion Inhibitors in Concrete. *Concrete International*, **13**(7) (1991), 24–7.

7.30. Berke, N.S. & Weil, T.G., World Wide Review of Corrosion Inhibitors in Concrete. In *Advances in Concrete Technology*, ed. V.M. Malhotra. CANMET, Ottawa, 2nd edition, 1994, pp. 891–914.

7.31. Berke, N.S. & Rosenberg, A., *Technical review of calcium nitrite corrosion inhibitors in concrete*. Transportation Research Record, No. 1211, Transportation Research Board, Washington, DC, 1989, pp. 18–27.

7.32. El-Jazairi, B. & Berke, N.S., The use of calcium nitrite as a corrosion inhibiting admixture to steel reinforcement in concrete. In *Corrosion of Reinforcement in Concrete*, eds C.L. Page, K.W.J. Treadaway & P.B. Bamforth. Society of Chemical Industry, Elsevier Applied Science, 1990, pp. 571–85.

7.33. Keer, J.G., Surface Treatments. In *Durability of Concrete Structures*, ed. G. Mays. E & FN SPON, UK, 1992, pp. 146–65.

7.34. CIRIA, Technical Note 130. Construction Industry Research and Information Association (CIRIA), London, 1987.

7.35. Weyers, R.E., Al-Qadi, I.L., Prowell, B.D., Dutta, T., Gouru, H. & Berke, N., Corrosion Protection System, Report to the Strategic Highway Research Program, SHRP Contract C-103, National Research Council, Washington, DC, USA, 1992.

7.36. Robinson, H.L., Durability of anti-carbonation coating. *J. Oil Color Chemists Association*, **70** (1987), 193–8.

7.37. Klopfer, H. Carbonation of exposed concrete and its remedies. *Bautenschutz Bausaneriung*, **1**(3) (1978), 86–96.

7.38. Concrete Society, Permeability testing of site concrete – a review of methods and experience. Report of a working party, Technical Report 31, Concrete Society, UK, 1988.

7.39. ACI Committee 515, Guide to the use of waterproofing, damp proofing, protective and decorative barrier systems for concrete. Manual of Concrete Practice, Part 5, American Concrete Institute, Detroit, 1994.

7.40. National Cooperative Highway Research Program (NCHRP), Concrete Sealers for the protection of bridge structures. Report 244, Transportation Research Board, Washington, 1981.

7.41. Robinson, H.L., An evaluation of silane treated concrete. *J. Oil Color Chemists Assoc.*, **70** (1987), 163–72.

7.42. Pfeifer, D.W. & Scali, M.J., Concrete Sealers for Protection of Bridge Structures. National Cooperative Highway Research Program Report 244, Transportation Research Board, National Research Council, Washington, DC, 1981.

7.43. Chandler, K.A. & Bayliss, D.A., *Corrosion Protection of Steel Structures*. Elsevier Science Publishers, 1985.
7.44. Roetheli, B.E., Cox, G.K. & Littreal, W.B., *Metals and Alloys*, March, **3**(3), (1932), 73. Cited from Ref 7.45.
7.45. Andrade, M.C. & Macias, A., Galvanized Reinforcement in Concrete. In *Surface Coating-2*, eds A.D. Wilson, J.W. Nicholson & H.J. Prosser. Elsevier Applied Science, pp. 137–82.
7.46. Andrade, C., Holst, J.D. (convenor), Nurenberger, U., Whiteley, J.D. & Woodman, N. Protection Systems for Reinforcement. CEB Information Bulletin No. 211, Switzerland, 1992.
7.47. Zinc Coated Reinforcement for Concrete. BRE Digest 109, Building Research Establishment, 1969, United Kingdom.
7.48. Yeomans, S.R., Performance of black, galvanized and epoxy-coated reinforcing steels in chloride-contaminated concrete. *Corrosion*, 50(1) (1994), 72–81.
7.49. Clear, K.C. Time to Corrosion of Reinforced Steel in Concrete Slabs, Vol. 4, Galvanized Reinforcing Steel. Interim report, Federal Highway Administration Report No. FHWA/RP-80/028, FHA, US Department of Transportation, Washington, DC, 1981.
7.50. Pfeifer, D.W., Landgren, J.R. & Zoob, A., Protective Systems for New Prestressed and Substructure Concrete. Report No. FHWA/RP-86/193, Federal Highway Administration, Washington, DC, April, 1987.
7.51. Anon, The Use of Galvanized Reinforcement in Concrete. Current Practice Note 17, Concrete Institute of Australia, 1994.
7.52. Pike, R.G. *et al.*, Nonmetallic coatings for concrete reinforcing bars. *Public Roads*, **37**(5) (1972), 185–97.
7.53. Gustafson, D.P. & Neff, T.L., Epoxy-coated rebar: handle with care. *Concrete Construction*, **39**(4) (1994), 356–69.
7.54. Cairns, J., Design of concrete structures using fusion-bonded epoxy-coated reinforcement. *Proc. Inst. Civ. Engrg Structures and Buildings*, **4**(2) (1992), 93–102.
7.55. ACI Committee 318, Revisions to building code requirements for reinforced concrete. *Amer. Concrete Inst. Structures J.*, **85**(6) (1988), 645–74.
7.56. Clifton, J.R., Beeghley, H.F. & Mathey, R.G., Nonmetallic coatings for concrete reinforcing bars. Building Science Series 65, US Department of Commerce, National; Bureau of Standards, August, 1975.
7.57. Clear, K. & Virmani, Y., Corrosion of nonspecification epoxy-coated rebars in salty concrete. *Public Roads*, **47**, June 1983.
7.58. Satake, J. Kamakura, M., Shirakawa, K, Mikami, N. & Swamy, R.N., Long term resistance of epoxy-coated reinforcing bars. In *Corrosion of Reinforcement in Concrete Construction*, ed. A.P. Crane. The Society of Chemical Industry/Ellis Horwood Ltd, UK, 1983, pp. 357–77.
7.59. Clear, K.C., Effectiveness of epoxy-coated reinforcing steel. *Concrete International*, **14**(5) (1992), 58–64.
7.60. Smith, L.L., Kessler, R.J. & Powers, R.G., Corrosion of epoxy coated rebar in a marine environment. Transportation Research Circular No. 403, Transportation Research Board, National Research Council, 1993, pp. 36–45.
7.61. Epoxy Coated Reinforcement in Highway Structures. Transportation Research Circular No. 403, Transportation Research Board, National Research Council, Washington, DC, March 1993.
7.62. RILEM Committee 124-SRC, Draft recommendation for repair strategies for concrete structures damaged by steel corrosion. *Materials and Structures*, **27**(171) (1994), 415–36.
7.63. Mays, G., *Durability of Concrete Structures*. E & FN SPON, UK, 1992.
7.64. Broomfield, J., Field Survey of Cathodic Protection on North American Bridges. *Materials Performance*, **31**(9) (1992), 28–33.

CHAPTER 8

# Specifications and Design

## 8.1. SPECIFICATIONS AND CODES

Most existing specifications designed to minimize the probability of steel corrosion in concrete deal with three specific parameters:

(1) The maximum level of chlorides to be permitted in the concrete.
(2) The thickness, composition and properties of the concrete cover.
(3) The maximum allowable crack width, as predicted from the design characteristics of the member concerned.

Many specifications deal separately with different types of concrete, that is, conventionally reinforced concrete and either cast in place or precast prestressed concretes; often the specific exposure conditions expected are also taken into account.

The most severe exposures are usually those of concretes in pavements and in marine structures. Such concretes are often in contact with chlorides. Salt is often used to facilitate snow melting on pavements in winter climates, and sea water contains a relatively high concentration of dissolved chloride salts. Reinforced concretes exposed to such environments can have a relatively short corrosion initiation stage, i.e. the time to depassivation can be minimal, especially where necessary precautions are not exercised.

In recent years, offshore structures constructed for oil exploration and production activities have been a particular focus of attention. In such structures, as in any marine structure, the greatest risk of steel corrosion occurs in portions extending above the level of the sea. Zones at some depth below the water surface, with reduced levels of diffusion of oxygen, are often anodic to the surface and are also at risk. Water spray results in accumulation and penetration of large amounts of chlorides into the concrete. In addition, freezing and thawing, wetting and drying, and shrinkage, and other environmental effects are much more severe than in the submerged portions of the structure. Cracking of the cover due to such causes further increases the likelihood of corrosion damage.

### 8.1.1. Maximum Chloride Levels Permitted in Concrete

The desirability of limiting the concentration of chloride ions within reinforced concrete was discussed in section 4.2, and the limiting chloride levels suggested by the relevant ACI specifications were presented in Table 4.1. Possible sources of chlorides in concrete-making materials include admixtures, contaminated aggregates, and contaminated water. The chloride level of commercial cements is ordinarily kept low enough that chloride from the cement itself should not be a problem.

The 1992 European pre-standard ENV 206, Concrete: Performance, Production Placing and Compliance Criteria [8.1], limits the chloride ion content introduced by the concrete ingredients to values of 1%, 0.4% and 0.2% by weight of cement for plain, reinforced and prestressed concrete, respectively. Furthermore, it forbids the use of calcium chloride and chloride-based admixtures in reinforced concrete or prestressed concrete.

The ACI documents present a variety of limitations of the contents of chloride. There is a limitation on the content of soluble chloride ions (i.e. water-soluble chlorides) [8.2], or on total chloride ions (i.e. acid-soluble chlorides) [8.3, 8.4] that are contributed from the concrete ingredients, as well as total content of chloride ions in the concrete resulting from penetration of chlorides during service [8.5]. These requirements and recommendations are summarized in Table 8.1. ACI Committee 357 [8.4] recommends that the mixing water used should not contain more than 0.07% chloride for conventionally reinforced concrete, and 0.04% chloride for prestressed concrete.

### 8.1.2. Thickness and Composition of Concrete Cover

#### *8.1.2.1. British Codes*

The effectiveness of the concrete cover as a barrier to corrosion depends both on its quality and on its thickness. High-quality concrete cover, that is, cover composed of uncracked, well-consolidated, low w/c ratio concrete is ordinarily significantly less permeable and more effective in preventing the entry of deleterious substances than concrete of lesser quality. Thus, the logical approach is to specify the required depth of cover as a function of its expected quality, a lesser depth being needed for high-quality cover.

The British Standard, BS 8110 [8.6] takes this approach. Details are presented in Table 8.2, as separately specified for the different exposure conditions contemplated and for the different qualities of cover. The latter are specified simultaneously in terms of maximum w/c ratio, minimum cement content and strength grades (in MPa).

**Table 8.1.** Requirements and Recommendations for Maximum Chloride Ion Content for Corrosion Protection set by Various ACI Documents.

| *Type of member* | *Maximum chloride ion content (% wt of cement)* | | | |
|---|---|---|---|---|
| | *Soluble*[a] | *Total*[b] | *Total*[c] | *Total*[d] |
| | *(in water)* | | *(in acid)* | |
| Prestressed concrete | 0.06 | | 0.06 | 0.08 |
| Reinforced concrete exposed to chloride in service | 0.15 | 0.10 | 0.10 | 0.20 |
| Reinforced concrete that will be dry or protected from moisture in service | 1.00 | | | |
| Other reinforced concrete construction | 0.30 | 0.15 | | |

[a]Chloride contributed from concrete ingredients; ACI Building Code (Ref. 8.2).
[b]Chloride contributed from concrete ingredients; ACI Guide to Durable Concrete (Ref. 8.3).
[c]Chloride contributed from concrete ingredients; ACI Guide for Design of Off-Shore Structures (Ref. 8.4).
[d]Limit of total chlorides in concrete (including chloride that penetrated during service) to minimize corrosion risks; ACI Corrosion of Metals in Concrete (Ref. 8.5).

#### *8.1.2.2. American Codes*

The ACI codes do not provide such detailed recommendations. They set requirements for minimum cover depth of cast in place concrete (Table 8.3) and provide recommendations for the minimum quality of concrete. In the ACI Guide to Durable Concrete [8.3] a maximum of 0.40 w/c ratio in structures exposed to sea water or salt spray is recommended. For structures exposed to salt spray it is also recommended that the concrete cover thickness be at least 50 mm. Because of construction tolerances it is suggested to set a design cover thickness of 65 mm to achieve the 50 mm minimum.

Remark: smaller cover thickness than specified in Table 8.3 is allowed in precast concrete components, reflecting the superior production control conditions (8.2).

#### *8.1.2.3. European Codes*

The European documents provide more detailed recommendations and specify different grades of concretes for different exposure conditions.

The 1992 European pre-standard, ENV-206 [8.1] (Concrete: Performance, Production, Placing and Compliance Criteria), the CEB guide [8.8] (Durable Concrete Structures: Design Guide) and the

**Table 8.2.** Specified Properties of Concrete Cover and Thickness According to BS 8110.[a]

| | | *Nominal cover thickness, mn* | | | | |
|---|---|---|---|---|---|---|
| | *Concrete grade*[b] | *C30* | *C35* | *C40* | *C45* | *C50* |
| *Exposure conditions* | *Max. w/c ratio*[c] | *0.65* | *0.60* | *0.55* | *0.50* | *0.45* |
| | *Min. cement*[d] | *275* | *300* | *325* | *350* | *400* |
| *Moderate*: concrete surfaces protected against weather or aggressive conditions | | 25 | 20 | 20 | 20 | 20 |
| *Moderate*: concrete surfaces sheltered from severe rain or freezing while wet; concrete subjected to condensation; concrete surfaces continuously under water; concrete in contact with non-aggressive soil | | – | 35 | 30 | 25 | 20 |
| *Severe*: concrete surfaces exposed to severe rain, alternate wetting and drying or occassional freezing or severe condensation | | – | – | 40 | 30 | 25 |
| *Very severe*: concrete surfaces exposed to sea water spray, de-icing salts (directly or indirectly), corrosive fumes or severe freezing conditions while wet | | – | – | 50 | 40 | 30 |
| *Extreme*: concrete surfaces exposed to abrasive conditions; e.g. sea water carrying solids or flowing water with pH ≤4.5 or machinery or vehicles | | – | – | – | 60 | 50 |

[a]From Ref. 8.6.
[b]Strength grade (MPa).
[c]Maximum w/c ratio.
[d]Minimum cement content (kg/m$^3$).

Eurocode 2 [8.9] specify maximum w/c ratio and minimum cement content (Table 8.4) and minimum concrete cover thickness. The 1993 British Standards document PD 6534: 1993 (Guide to the use in the UK of DD ENV 206: 1992 Concrete) [8.10], provides a guide for the minimum grades of concrete which generally ensure that the limits on w/c ratio and cement content of ENV 206 (Table 8.4). These values are presented in the last column in Table 8.4 in terms of minimum characteristic strength in MPa units ('characteristic' concrete strength – the value of strength below which 5% of the population of all possible strength measurements of the

**Table 8.3.** Minimum Thickness of Cover for Cast-In-Place Concrete As Specified According to the ACI Committee 301.[a]

| *Type of strucuture* | *Minimum cover (mm; (in))* |
|---|---|
| Concrete deposited against the ground | 75 (3) |
| Formed surfaces exposed to weather or in contact with ground: | |
| No. 6 bar or greater | 50 (2) |
| No. 5 bar or smaller | 38 (1.5) |
| Formed surface not exposed to weather or not in contact with the ground: | |
| Beams girders and columns | 38 (1.5) |
| Slabs, walls and joists, No. 11 bar or smaller | 19 (0.75) |
| Slabs, walls and joists, No. 14 and 18 bars | 38 (1.5) |

[a]From Ref. 8.7.
Comment: smaller cover thickness than specified in Table 8.3 is allowed in precast concrete components, reflecting the superior production control conditions (Ref. 8.2).

specified concrete are expected to fall). The first numerical value in this column is the cube characteristic strength and the second one after the slash is the corresponding cylinder characteristic strength. Also, the British Guide PD 6534 [8.10] points out that the exposure classes in ENV 206 (Table 8.4) and BS 8110 (Table 8.2) and BS 5328 (Table 8.4) can not be directly related, although there are obvious similarities.

Barker and Mathews [8.11] compared between the European and British standards and suggested that the British standard is more adequate when considering the use of blended cements. They recommended that blended cement concretes should be preferably specified on the basis of concrete strength level rather than w/c ratio (see discussion in section 7.2.2).

It should be kept in mind that the thicknesses of cover specified are nominal thicknesses. Improper or casual construction practices may result in large variations in the depth of cover actually obtained, with a significant proportion of the steel sometimes being placed with cover thicknesses significantly less than specified. Browne *et al.* [8.12] discussed an extreme case in which the average depth of cover, 13.9 mm, was only about half of the nominal design value of 25 mm. Van Daveer [8.13] reported results of a survey of concrete cover in bridge decks in which the nominal design thickness was 38 mm (1½ in); however, the actual standard deviation shown was as high as 9.5 mm ($^3/_8$ in). If, as suggested by these authors, a cover depth of 50 mm (2 in) was actually required on a bridge deck, it would be necessary to specify a nominal value of 70 mm ($2^3/_4$ in).

Eurocode 2 [8.9] specifies that the requirement for minimum cover thickness (Table 8.4) should be increased by an allowance $\Delta h$ for tolerances. Values for $\Delta h$ should be in the range of 0 to 5 mm for precast

**Table 8.4.** Concrete requirements for durability according to the European pre-standard ENV-206 (Ref. 8.1) and additional recommendation according to the British Standard Guide DD ENV 206 (Ref. 8.11)

| Exposure class | | Max w/c ratio[a] | | Min cement content ($kg/m^3$)[a] | | Min concrete[c] cover (mm) | | Concrete grade[c] |
|---|---|---|---|---|---|---|---|---|
| | | Reinf. | Prestressed | Reinf. | Prestressed | Reinf. | Prestressed | |
| 1. Dry | | 0.65 | 0.60 | 260 | 300 | 15 | 25 | C30/37 |
| 2. Humid | a) no frost | 0.60 | 0.60 | 280 | 300 | 20 | 30 | C30/37 |
| | b) frost | 0.55 | 0.55 | 280 | 300 | 25 | 35 | C35/45 |
| 3. 2b + De-icing salts | | 0.50 | 0.50 | 300 | 300 | 40 | 50 | C35/45 |
| 4. Sea water | a) no frost | 0.55 | 0.55 | 300 | 300 | 40 | 50 | C35/45 |
| | b) frost | 0.50 | 0.50 | 300 | 300 | 40 | 50 | C35/45 |
| 5. Aggressive chemical[d] | a) slightly | 0.55 | 0.55 | 280 | 300 | 25 | 35 | – |
| | b) moderately | 0.50 | 0.50 | 300 | 300 | 30 | 40 | – |
| | c) highly[e] | 0.45 | 0.45 | 300 | 300 | 40 | 50 | – |

[a] According to European pre-standard ENV 206 (Ref. 8.1).
[b] According to Eurocode 2: Design of Concrete Structures (Ref. 8.9).
[c] Recommendation according to British Standard Guide PD 6534 for the use of the European pre-standard 206 (Ref. 8.10); the letter C denotes characteristic strength grade, the following two digits the characteristic cube strength, and the two digits after the slash the cylinder characteristic strength.
[d] Sulphate-resistant cement if sulphate >500 mg/kg in water or >3000 mg/kg in soil.
[e] Additional coating over the concrete is required.

elements if production is adequately controlled, and 5–10 mm for *in situ*-reinforced concrete production. This approach is also taken by other specifying agencies like ACI [8.2].

For prestressed concrete, however, most specifications require larger depths of cover because of the greater damage that may accompany the corrosion of highly stressed steel.

#### *8.1.2.4. Specifications for Special Structures and Severe Exposure Conditions*

The most stringent requirements are often specified for marine concrete, where reference is often made not only to the thickness of the cover, but also to the composition and properties of the concrete. As an example, the requirement of ACI Committee 357 on Design and Construction of Fixed Concrete Offshore Structures [8.4] are shown in Table 8.5. It is recommended by this committee that concrete with a minimum cement content of 360 kg/m$^3$ (600 lb/yd$^3$) always be used for such structures to ensure the quality of the paste adjacent to the reinforcement.

The requirements for the splash zone of marine structures specified by several different agencies were discussed by Browne and Baker [8.14], Browne [8.15] and Marshal [8.16], and are summarized and compared in Table 8.6 below.

It would seem that a test of the permeability of the finished concrete cover would provide an extremely valuable parameter for quality control purposes, since permeability is directly related to the effectiveness of the cover as a protective coating. However, due to the difficulty of carrying out permeability tests on a routine basis with high-quality concrete, only the DNV recommendations call for permeability tests to be made.

A compilation of the main requirements of the British Standards and Codes for severe exposure conditions and special structures was recently presented by Bamforth [8.17], and is presented here in Table 8.7. The

**Table 8.5.** ACI Committee 357 Recommendation for Concrete Quality and Depth of Cover in Offshore Structures.[a]

| *Zone* | *Maximum w/c ratio* | *Minimum 28-day compressive strength (Mpa; (psi))* | *Depth of cover*[b] | |
|---|---|---|---|---|
| | | | *Conventional steel (mm; (in))* | *Post-tensioned ducts (mm; (in))* |
| Atmospheric zone[c] | 0.40 | 35 (5000) | 50 (2) | 75 (3) |
| Splash zone | 0.40 | 35 (5000) | 65 (2.5) | 90 (3.5) |
| Submerged zone | 0.45 | 35 (5000) | 50 (2) | 75 (3) |

[a] From Ref. 8.4.
[b] Cover of stirrups may be 13 mm (0.5 in) less than the value cited below.
[c] Above splash zone.

table includes also Bamforth's estimates for the effective chloride diffusion coefficients of the concretes and calculated service life. The latter information will be further discussed in section 8.2

In conclusion of this section, it should be emphasized that the quality of a concrete cover is not solely a function of its w/c ratio, cement content and strength grade, even though these are almost the only quality parameters specified. The attainment of a workable mix in practice, effective consolidation, and effective curing are also essential. Proper aggregate grading and specification of maximum size of aggregate need to be carefully controlled to produce a mix that can be properly compacted around the reinforcing bars.

### 8.1.3. Maximum Crack Width

The effect of cracks in influencing the corrosion processes in structures with cracked concrete cover was discussed in section 4.3.2, and the maximum allowable loading crack width recommendations provided by ACI were indicated in Table 4.4. Additional maximum allowable crack width specifications for marine concretes were listed in Table 8.6.

In designing a concrete structure against corrosion, the several specifications provide equations to predict the maximum crack widths and crack spacing that are expected to result from the loads and stresses carried by the structure. ACI Committee 224 [8.18] suggested the following equation for the maximum crack width:

$$W_{max} = 0.076\,\beta\,f_s\,(d_c A)^{1/3} \times 10^{-3} \tag{8.1}$$

**Table 8.6.** Comparison of Concrete Design Specifications for the Splash Zone in Marine Environments.[a]

| *Agency* | *Depth of cover (mm)* | *Max. crack width (mm)* | *Max. w/c ratio* | *Min. cement content (kg/m³)* | *Permeability coefficient (m/s)* |
|---|---|---|---|---|---|
| DNV[b] | 50 | – | 0.45 | 400 | $10^{-12}$ |
| FIP[c] | 75 | 0.004 × cover, or 0.3 | 0.45 | 400 | – |
| BS 6235[d] | 75 | 0.004 × cover, or 0.3 | 0.40 | 400 | – |
| ACI[e] | 65 | – | 0.40 | 360 | – |

[a] Adapted from Refs 8.14–8.16.
[b] Det Norske Veritas.
[c] Federation of Prestressed Concrete.
[d] British Standard.
[e] American Concrete Institute.

**Table 8.7.** Current British Standard Code Requirement, Predicted Values of $D_{eff}$ and Estimated Times both to the Initiation of Corrosion ($C_i = 0.4\%$ by wt. of cement) and to Onset of Rapid Corrosion ($C_p = 1.0\%$ by weight of cement).[a]

| | | | Concrete mix details | | | | | Service life (years) | |
|---|---|---|---|---|---|---|---|---|---|
| *Source* | *Exposure class* | *Exposure condition* | *Min. Cement* $(kg/m^{-3})$ | *Max. w/c* | *Min. Grade (mm)* | *Min. cover (mm)* | *Effective diffusion coefficient* $(m^2/s)$ | $C_i = 0.4\%$ | $C_p = 1.0\%$ |
| General structures BS 8110 | Very severe | Sea water spray, de-icing salts or severe freezing conditions when wet | 325 | 0.55 | 40 | 50 | $3.93 \times 10^{-12}$ | 3.1 | 5.6 |
| | | | 350 | 0.50 | 45 | 40 | $3.18 \times 10^{-12}$ | 2.6 | 4.6 |
| | | | 400 | 0.45 | 50 | 30 | $2.57 \times 10^{-12}$ | 1.9 | 3.7 |
| | Extreme | Abrasive action, sea water carrying solids | 350 | 0.50 | 45 | 60 | $3.18 \times 10^{-12}$ | 5.8 | 10.4 |
| | | | 400 | 0.45 | 50 | 50 | $2.57 \times 10^{-12}$ | 5.4 | 10.2 |
| Bridges BS 5400 Part 4 | Very severe | Direct de-icing salts or sea water spray | 360 | ns | 40 | 50 | $3.93 \times 10^{-12}$ | 3.3 | 6.1 |
| | | | or 330 | 0.45 | 50 | 40 | $2.57 \times 10^{-12}$ | 3.0 | 5.5 |
| | Extreme | Abrasive action by sea water | 360 | ns | 40 | 65 | $3.93 \times 10^{-12}$ | 5.5 | 10.3 |
| | | | or 330 | 0.45 | 50 | 55 | $2.57 \times 10^{-12}$ | 5.7 | 10.3 |
| Maritime structures BS 6349 Part 1 | Continuously submerged | In sea water up to a level of 1 m. below LWL | 350 | 0.50 | ns | >50 but | $3.18 \times 10^{-12}$ | 3.3 | 7.2 |
| | | | | | | 75 pref. | $3.18 \times 10^{-12}$ | 7.4 | 16.2 |
| | Tidal/splash | Spray or tidal zone to 1 m. below LWL, or severe abrasion | 400 | 0.45 | ns | >50 but | $2.57 \times 10^{-12}$ | 5.4 | 10.2 |
| | | | | | | 75 pref. | $2.57 \times 10^{-12}$ | 12.0 | 22.9 |
| ENV 206 | Sea water XS1 | Airborne salt | 330 | 0.50 | 40 | 35 | $3.93 \times 10^{-12}$ | 1.5 | 2.7 |
| | XS2 | Submerged | 330 | 0.50 | 40 | 40 | $3.93 \times 10^{-12}$ | 2.0 | 3.6 |
| | XS3 | Tidal, splash, spray | 350 | 0.45 | 45 | 40 | $3.18 \times 10^{-12}$ | 2.1 | 4.6 |
| | Chlorides other than sea water XS4 | Wet, rarely dry | 300 | 0.55 | 40 | 40 | $3.93 \times 10^{-12}$ | 2.0 | 3.4 |
| | XS5 | Cyclic wet and dry | 330 | 0.50 | 40 | 40 | $3.93 \times 10^{-12}$ | 2.0 | 3.6 |

where $W_{max}$ is the maximum probable crack width in inches, β is the ratio of distance between the neutral axis and the tension face to the distance between the neutral axis and the centroid of the reinforcing steel, $f_s$ is the reinforcing steel stress in k.s.i. ($10^3$ p.s.i.), $d_c$ is the thickness of cover in inches, and $A$ is the area of concrete symmetric with the reinforcing steel in $in^2$ divided by the number of bars.

Nominal crack width prediction equations such as eqn. (7.2) were recently discussed and compared by Beeby [8.19]. Load-induced cracks are obviously not the only type of cover cracking that might develop in a given structure. Cracking may also occur from repeated freezing and thawing, plastic shrinkage, drying shrinkage, and various chemical reactions such as sulphate attack and alkali–aggregate reactions. In principle, such cracks can be eliminated by proper attention to the concrete components employed and to the design of the concrete mix. In practice, this is not necessarily true, and corrosion is often found as a secondary consequence of, and associated with, cracking produced by these other environmental factors. Needless to say, it is not possible to predict the widths and spacings of such cracks in the same way that loading cracks can be predicted.

The effects of cracks in concrete cover may vary not only with their effective widths, but also with their orientation. It is generally accepted that cracks parallel to the primary reinforcement are much more critical than transverse cracks, since much more of the steel surface is exposed. Such parallel cracks are usually produced by rust accumulation within the cover over the steel. The consequent exposure of large steel areas often results in a rapid increase in the corrosion rate, as was shown schematically in Fig. 4.2.

It is also possible that cracks can develop over or parallel to transverse or secondary reinforcing bars, thus being oriented transversely to the primary reinforcing steel. Such cracking was discussed by Beeby [8.19]. It is illustrated schematically in Fig. 8.1, for the case where the crack happens to form just above a transverse bar. The subsequent corrosion of this bar and the consequent reduction in its cross-sectional area may not cause a direct safety risk, since the bar is not stressed, and it does not play a major role in the stability of the structure. Nevertheless, as indicated by Beeby, if spalling occurs due to this corrosion, the main reinforcement can subsequently be exposed and its corrosion accelerated. This consequence emphasizes the necessity for the provision of adequate corrosion protection for the transverse bars and stirrups as well as for the primary reinforcement.

## 8.2. DESIGN CONSIDERATIONS FOR CARBONATION-INDUCED CORROSION

The carbonation of concrete can be quantified by the diffusion equations (see section 4.3.1.1) or by empirical relations such as eqn. (4.4) in section

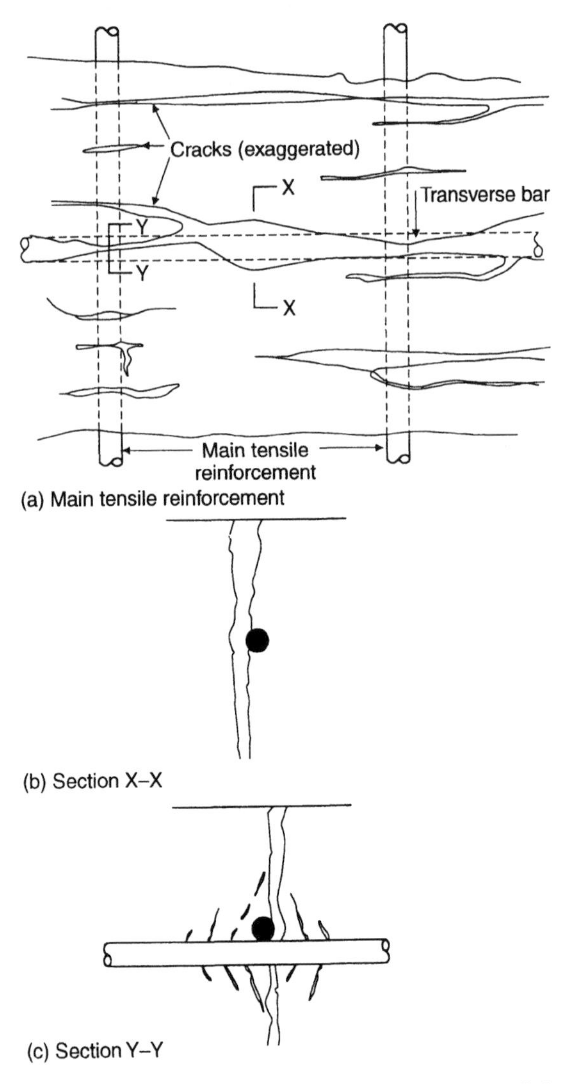


**Fig. 8.1.** Schematic description of the formation of cracks parallel to the secondary reinforcement. (After Ref. 8.19)

4.3.1.4. These equations can be used to predict the time required for the carbonation front to reach the steel level. This is the time $t_i$ required for the initiation of the corrosion (see section 4.2). On top of that, additional time, $t_p$, can elapse in the propagation stage before sufficient steel corrosion has occurred to cause cracking or spalling in the concrete cover. Estimation of $t_i + t_p$ is the quantitative basis for the design to control the carbonation-induced corrosion.

Equation (4.4) can be used to estimate $t_i$ by introducing experimentally determined coefficients to the values of $n$ and $k$. $n$ is usually about 2 and $k$

can vary considerably depending on the type of concrete and exposure conditions. The $k$ values can be determined on the basis of experience or natural exposure tests and on the basis of accelerated carbonation tests. Linear correlations were reported between $k$ determined in accelerated tests (which are conducted in artificially rich $CO_2$ environment) and in natural exposures [8.20]. An example is presented in Fig. 8.2. Using such data $t_i$ can be calculated as:

$$t_i = (D/k)^{1/2} \tag{8.2}$$

where $k$ is the empirical coefficient for particular exposure conditions and concrete quality; and $D$ is concrete cover thickness.

Parrott [8.21] has recently suggested a more comprehensive approach to estimate $t_i$ and $t_p$. For the calculation of $t_i$ the following assumptions were made:

- The coefficient of $CO_2$ penetration can be estimated from the air permeability of the cover concrete.
- The binding capacity towards the diffusing $CO_2$ can be related to the CaO content of the cement and its degree of hydration.
- Variations in the atmospheric $CO_2$ content can be ignored.
- Under wetter conditions there is a departure from the square root time functions predicted by equations such as (8.2).

On the basis of these concepts, Parrott suggested the following equation:

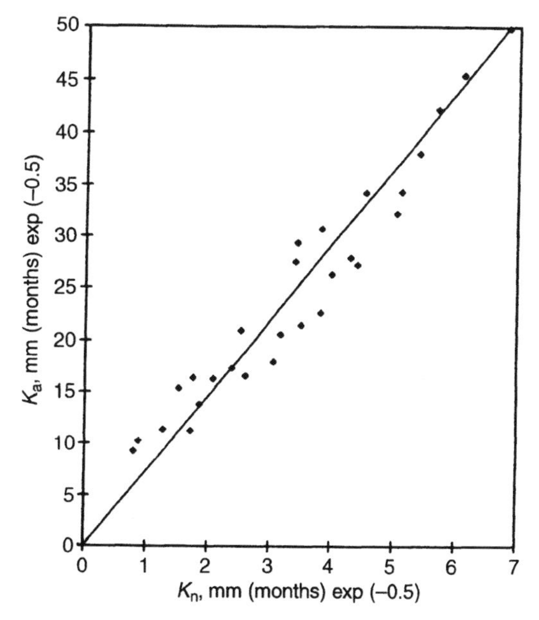


**Fig. 8.2.** Relations between the $k$ constants in eqn (4.4) determined by accelerated tests, Ka, and natural exposure, Kn. (After Ref. 8.20)

$$D = a \bullet k^{0.4} . t_i^{n} / c^{0.5} \quad (8.3)$$

where:

- $D$ is depth of carbonation; and $k$ is air permeability (in units of $10^{-16}$ $m^2$) of the concrete cover. This value is dependent on relative humidity, and can be estimated from the value of $k$ at 60% RH, $k_{60}$, using the relationship: $k = m \bullet k_{60}$. Values of $m$ for different relative humidities are provided in Table 8.8.
- $n$ is a power exponent which is usually about 0.5, but decreases as the relative humidity increases above 70% RH; values of $n$ as a function of relative humidity are given in Table 8.8.
- $c$ is the CaO content in the concrete cover. Estimates for different types of European cements and different relative humidities (which control extent of hydration) are given in Table 8.8 (in units of kg/m³ of cement matrix); these values would be sensitive to local conditions.
- $a$ is a coefficient which can be assigned the value of 64 based on available data.

Estimates of the propagation period $t_p$ were established by Parrott based on the following equation:

$$t_p = CD/CR \quad (8.4)$$

where CD is corrosion depth of steel required to cause cracking in the concrete cover; and CR is corrosion rate of the steel (in units of steel depth per year) in the propagation stage.

CD was estimated to be 0.1 mm (see section 5.1) while CR can be estimated for different relative humidities as shown in Table 8.8. The latter values are rough estimates based on data in the literature and are sensitive to a variety of influences. The values chosen by Parrott in Table 8.8 are such so that they have about 10% chance of being exceeded.

**Table 8.8.** Values of $m$, $n$, $c$ (kg/m³), CR (μm/y) and $t_p$ (years) as a function of relative humidities to be used in eqns (8.3), (8.4) and (8.5).[a]

| *Relative humidity (%)* | *40* | *50* | *60* | *70* | *80* | *90* | *95* | *98* | *100* |
|---|---|---|---|---|---|---|---|---|---|
| $m$ | 1.00 | 1.00 | 1.00 | 0.797 | 0.564 | 0.301 | 0.160 | 0.071 | 0.010 |
| $n$ | 0.480 | 0.512 | 0.512 | 0.480 | 0.415 | 0.317 | 0.256 | 0.216 | 0.187 |
| $c$ CEM1[b] | 460 | 460 | 460 | 460 | 485 | 535 | 570 | 595 | 610 |
| CEM2[b] | 360 | 360 | 360 | 360 | 380 | 420 | 445 | 465 | 480 |
| CEM3[b] | 340 | 340 | 340 | 340 | 355 | 395 | 420 | 440 | 450 |
| CEM4[b] | 230 | 230 | 230 | 230 | 240 | 265 | 285 | 295 | 305 |
| CR | 0.3 | 0.3 | 0.3 | 2 | 5 | 10 | 20 | 50 | 10 |
| $t_p$ | 330 | 330 | 330 | 50 | 20 | 10 | 5 | 2 | 10 |

[a] After Ref. 8.21.
[b] Cement types according to European standards.

Based on eqns (8.3) and (8.4), the relationship between the depth of concrete cover and the service life $t = t_i + t_p$ was calculated:

$$D = 64 \bullet k^{0.4} \bullet [t-(100/CR)]^n / c^{0.5} \qquad (8.5)$$

Using eqn. (8.5) and empirical relations between the air permeability and the effective concrete strength, Parrott calculated the required concrete quality (i.e. strength grade and curing time) and concrete cover thickness to achieve a specified service life. The relations between the 28 days compressive strength, $f_{28}$, the effective concrete strength, $f_t$ (for a curing period other than 28 days) and the air permeability, $k_{60}$, were as follows:

$$f_t = f_{28} \bullet [0.25 + 0.225 \log(t)] \qquad (8.6)$$

$$\log(k_{60}) = (30-f_t)/10 \qquad (8.7)$$

where $t$ is the curing period + 8 days.

Results of calculations for a service life of 75 years for four exposure conditions as specified by the European prestandard ENV206: dry, moderate, drying/wetting and wet are presented in Table 8.9. The calculated concrete cover parameters were similar to those required by ENV 206, except for the wet conditions where the prestandard ENV 206 required higher quality concrete than that calculated (Table 8.9).

It is of interest to note that the data in Table 8.9 suggests that a minimum concrete grade of 30 MPa and a nominal concrete cover thickness of 30 mm will assure adequate performance even in the harshest exposure conditions.

**Table 8.9.** Concrete Qualities Required for Various Exposure Conditions, calculated ('calculated') from eqn (8.5) for 75 years service-life and compared with requirements specified ('specified') by European prestandard ENV 206.[a]

| *Exposure conditions* | *Nominal cover (mm)* | *Minimum cover (mm)* | $k_{60}$ *($10^{-16}$ m²)* | k *($10^{-16}$ m²)* | *Basis* | *Grade cube (MPa)* | *Water/ cement ratio* | *Relative humidity (%)* |
|---|---|---|---|---|---|---|---|---|
| Dry | 25 | 15 | 36 | 42 | calculated | 16 | 0.82 | 50 |
| | 25 | 15 | – | – | specified | 20 | 0.90 | – |
| Moderate | 25 | 15 | 46 | 42 | calculated | 15 | 0.85 | 65 |
| | 25 | 15 | 1.1 | 0.84 | calculated | 33 | 0.52 | 70 |
| | 25 | 15 | – | – | specified | 25 | 0.65 | – |
| Drying/ wetting | 30 | 20 | 1.7 | 1.2 | calculated | 31 | 0.55 | 50–100 |
| | 30 | 20 | 2.5 | 1.4 | calculated | 29 | 0.57 | 60–100 |
| | 30 | 20 | – | – | specified | 35 | 0.55 | – |
| Wet | 30 | 20 | 13 | 3.8 | calculated | 21 | 0.7 | 90 |
| | 30 | 20 | 46 | 7.4 | calculated | 15 | 0.85 | 95 |
| | 30 | 20 | – | – | specified | 30 | 0.60 | – |

[a]After Ref. 8.21.

## 8.3. DESIGN CONSIDERATIONS FOR CHLORIDE EXPOSURES

### 8.3.1. Introduction

In this section the effects of mixture design on the ingress of chloride as a function of environment and geometry of the component will be addressed. Chloride profiles will be calculated and will then be used to predict the service lives of structures. Several environments will be considered, ranging from parking and bridge decks in temperate climates to marine piles in locations with mean average temperatures above 25°C. It will be demonstrated that in most cases the minimal code requirements are not sufficient to provide the desired service lives, and that lower-permeability concretes with additional corrosion protection systems are necessary.

The design concepts are based on calculating chloride profiles at different times using eqn. (4.2). The calculation of the service life is shown schematically in Fig. 8.3. For its calculation three parameters are required:

(1) The effective diffusion coefficient, $D_{eff}$, which is mainly a function of the concrete quality.
(2) The effective surface concentration of the chlorides, C(o), which is mainly a function of the exposure conditions, but may be influenced slightly by the concrete composition.

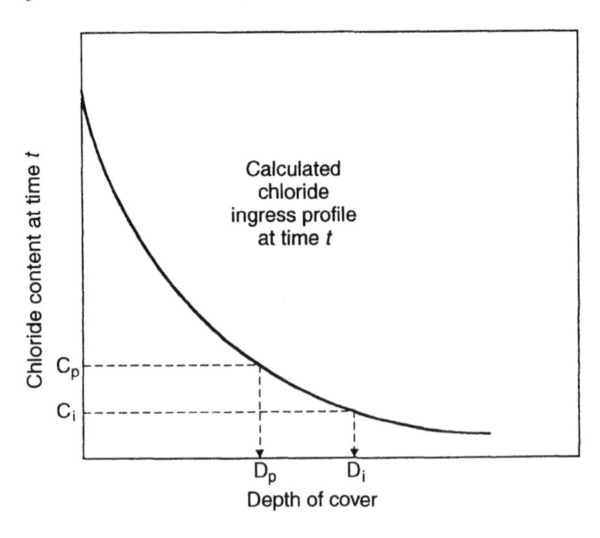


**Fig. 8.3.** Schematic description of the determination of the depth of concrete cover to achieve specified service life, *t*, based on the calculated chloride profile (using $D_{eff}$ for the particular concrete and $C_o$ for the exposure conditions) at time t. The threshold values $C_i$ and $C_p$ are for corrosion initiation and corrosion damage. Cover depth $D_i$ and $D_p$ are the ones required for corrosion initiation and corrosion damage, respectively.

(3) The threshold value of the chloride concentration required for corrosion to occur. Two threshold values have been used: one which is required for corrosion initiation, $C_i$ (point I in Fig. 4.2) and for damaging the concrete cover, i.e. cracking or delamination, $C_p$ (point P in Fig. 4.2).

For the design purposes, procedures and guidelines have been developed to estimate these three parameters. These procedures have been recently summarized by Bamforth [8.17] and Berke and Hicks [8.22], and they will be discussed here.

### 8.3.2. Effective Diffusion Coefficient, $D_{eff}$, for Chloride Ingress

The best method for determining $D_{eff}$ is by actual determination of the chloride content profile at a given time for concrete after severe chloride exposure for at least 2 years (or at least 1 year for concretes with w/c values above 0.4). The geometry should be fixed so that the chloride diffusion is one-dimensional, i.e. a flat surface should be exposed and no chloride should enter from the sides. In this case, non-linear regression analysis can be used to fit the data to eqn. (4.2), as demonstrated in Fig. 7.14. Solving of the equation will give the pseudo (or effective) surface concentration and the effective chloride diffusion coefficient. The surface effective concentration, C(o), will be in units of kg/m$^3$, or concentration as a percent weight of cement or concrete depending upon how the chloride data were entered. $D_{eff}$ is given in units of m$^2$/s. The surface concentration of chloride is dependent upon the exposure conditions, whereas $D_{eff}$ is a materials property of the concrete.

In many cases the designer neither has historical data on $D_{eff}$ for the concrete that is planned to be used, nor the time to determine the chloride content profile. For these situations, it is possible to estimate $D_{eff}$ from the conductivity or resistivity properties of the concrete [8.22]. This is feasible since the electrical resistivity is linearly related to the Coulomb value obtained in the rapid chloride permeability test (ASTM C1204-94) (Fig. 8.4) and both are linearly related to the effective diffusion coefficient, $D_{eff}$ (Fig. 8.5). In the absence of hard data these predictions give a reasonable estimation of $D_{eff}$ [8.22]. Equations quantifying these relations were presented by Berke and Hicks [8.22]:

$$D_{eff} = 54.6 \cdot 10^{-8} (\text{resistivity})^{-1.01} \qquad r^2 = 0.95 \qquad (8.8)$$

$$D_{eff} = 0.0103 \cdot 10^{-8} (\text{rapid permeability–Coloumb})^{-0.84} \qquad r^2 = 0.97 \qquad (8.9)$$

where the resistivity is in units of kohm • cm, rapid permeability in Coulombs and $D_{eff}$ in cm$^2$/s.

In many instances, $D_{eff}$ is determined at a temperature which is different than the one existing in the field. Equation (7.1) gives the relationship between the diffusion coefficient determined at one temperature (in units

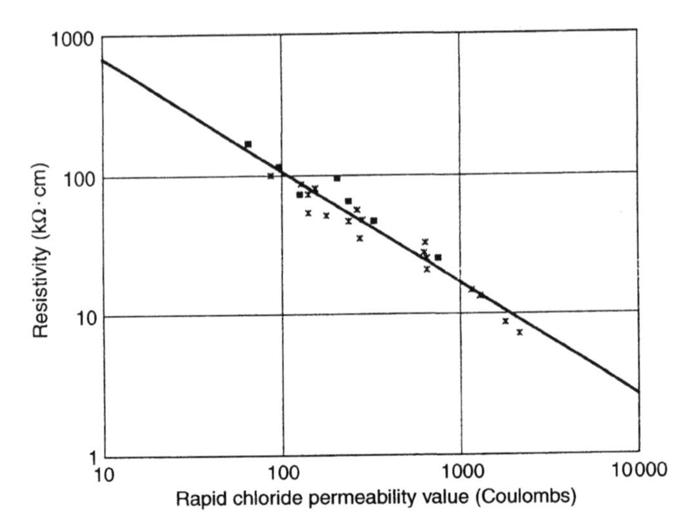


**Fig. 8.4.** Relationship between concrete electrical resistivity and the Coulomb value obtained in the rapid chloride permeability tests, ASTM C1204-94. (After Ref. 8.22)

of °K) and the diffusion coefficient at another temperature. The constant k in eqn. (7.1) is related to the activation of energy for chloride diffusion.

In the absence of any data on the actual concretes to be used, some general guidelines can be obtained from data compiled by Bamforth [8.17] and Berke and Hicks [8.22] in which $D_{eff}$ values were presented as a

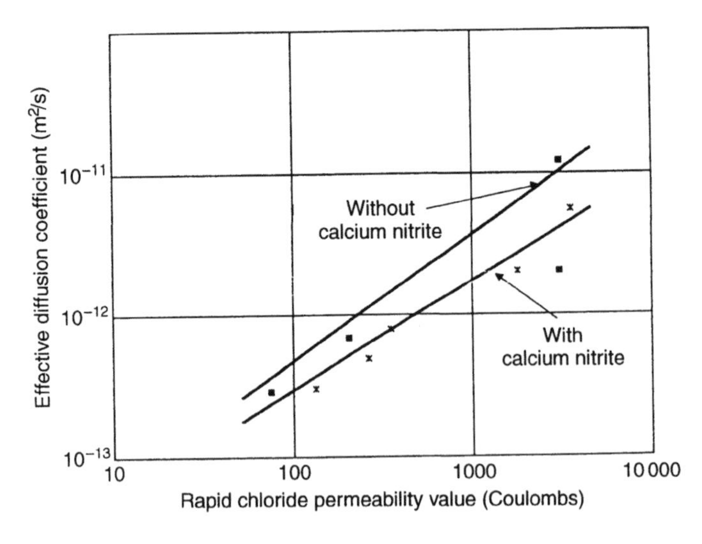


**Fig. 8.5.** Relationship between concrete Coulomb value obtained in the rapid chloride permeability test (ASTM C1204-94) and the effective diffusion coefficient, $D_{eff}$. (After Ref. 8.22)

function of the concrete quality. The approach taken by Bamforth [8.17] is to relate the effective diffusion to the concrete compressive strength (Fig. 8.7 and Table 8.7), whereas Berke and Hicks [8.22] presented it in terms of the w/c ratio (Table 8.10).

Table 8.10 includes also values calculated for different temperatures which are characteristic to the whole range typical for cold and warm climates. The values in Table 8.10 were determined from fitting eqn. (4.2) to actual chloride profile data, or from using eqns (8.8) and (8.9) obtained from the data in Figs 8.4 and 8.5. These values should be used as an estimate of $D_{eff}$, but not as an absolute value for a given concrete. The initial values of $D_{eff}$ were determined at 22°C (295°K). Examination of Table 8.10 shows that increasing temperature can significantly increase the rate of chloride ingress. Thus, designs in warmer climates require lower permeability concretes. This will be evident in the design examples given below.

For concretes made of Portland cement, only the effective diffusion values are hardly affected by age [8.17]. However, for concretes with mineral admixtures (fly-ash, blast furnace slag and silica fume) the w/c ratio and compressive strength can not serve as the overriding parameter for concrete quality, since with these admixtures the effective diffusion is reduced considerably relative to the value expected on the basis of the w/c ratio or strength, especially under prolonged exposures. The approach taken by Bamforth to account for these influences is to present special relations between $D_{eff}$ and compressive strength for concretes with mineral admixtures (Fig. 8.6). Berke and Hicks [8.22] chose another alternative in which the high-quality concretes with chemical admixtures

**Table 8.10.** Estimates of $D_{eff}$ as a Function of Concrete Mixture Proportions and Temperature.

| | $D_{eff}$ ($m^2/s$) | | | |
|---|---|---|---|---|
| *Mixture* | *22°C* | *10°C* | *18°C* | *27°C* |
| w/c = 0.50 | $12\times10^{-12}$ | $5.3\times10^{-12}$ | $9.2\times10^{-12}$ | $17\times10^{-12}$ |
| w/c = 0.45 | $6\times10^{-12}$ | $2.6\times10^{-12}$ | $4.6\times10^{-12}$ | $8.3\times10^{-12}$ |
| w/c = 0.40 | $3\times10^{-12}$ | $1.3\times10^{-12}$ | $2.3\times10^{-12}$ | $4.2\times10^{-12}$ |
| w/c = 0.35 or 0.4 with fly-ash | $2\times10^{-12}$ | $8.8\times10^{-13}$ | $1.5\times10^{-12}$ | $2.8\times10^{-12}$ |
| Equivalent to 1500 Coulombs | $2.6\times10^{-12}$ | $1.2\times10^{-12}$ | $2.0\times10^{-12}$ | $3.4\times10^{-12}$ |
| Equivalent to 1000 C[a] (typical to 5% SF) | $1.9\times10^{-12}$ | $8.3\times10^{-13}$ | $1.4\times10^{-12}$ | $2.6\times10^{-12}$ |
| Equivalent to 600 C[a] (typical to 7.5% SF) | $1.3\times10^{-12}$ | $5.7\times10^{-13}$ | $9.9\times10^{-13}$ | $1.8\times10^{-12}$ |
| Equivalent to 300 C[a] (typical to 15% SF) | $0.75\times10^{-12}$ | $3.3\times10^{-13}$ | $5.7\times10^{-13}$ | $1.04\times10^{-12}$ |

[a] Coulombs measured by ASTM C 1202-94 after 28 days moist care. These mixes will have various combinations of silica fume, fly-ash, or ground blast furnace slag, with w/b at 0.45 or lower. The curing should be increased to 90 days.

Note: the addition of calcium nitrite will cause an interference in the measurements of the Coloumb value according to ASTM C 1202-94 and the value for concrete without calcium nitrite should be used for estimates of $D_{eff}$.

were characterized in terms of their Coulomb value determined by the rapid chloride permeability test (ASTM C1202-94), as shown in Table 8.10. A 300-Coulomb value was included in Table 8.10. However, it is generally known that w/b values under 0.35 and silica fume contents above 10% are needed to meet this in the field. Experience suggests that field concretes with 7.5% silica fume and w/c <0.4 were needed to obtain a concrete with values smaller than 1000 Coulomb.

### 8.3.3. Effective Chloride Concentration at the Concrete Surface

Effective chloride concentration at the surface, C(o), can be determined by the technique described in section 8.3.2 whereby linear regression analysis is used to fit actual chloride profiles in existing structures to eqn. (4.2). The values are usually presented in units of % wt. of the concrete, or in units of kg/m³ of the concrete.

Berke and Hicks [8.22] reported that these values are mainly a function of the environmental conditions. Based on field surveys they provided the following recommendations to be used for design:

- severe marine exposure: C(o) quickly reaches a high constant value of 17.8 kg/m³.
- salt air exposure: C(o) increases at a rate of 0.15 kg/m³/year, up to a maximum of 15 kg/m³.

Bamforth [8.17] highlighted the influence of the concrete composition and surface properties on C(o). If the concrete is not fully saturated when

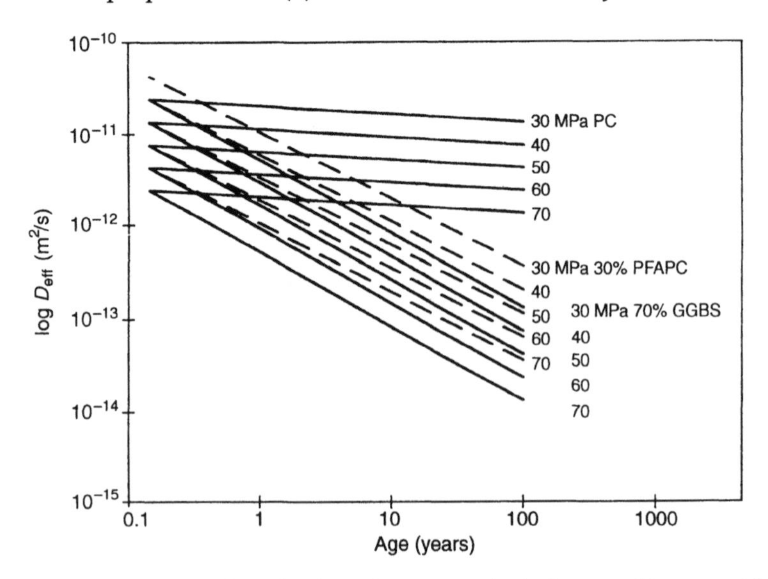


**Fig. 8.6.** Design curves for selection of concrete mix design to achieve specified values of $D_{eff}$. (After Ref. 8.17)

first exposed to salt the absorption capacity – as quantified by the sorptivity value – will affect C(o), as shown in Fig. 8.7. Concretes with higher cement contents were found to have higher C(o) (Fig. 8.8)and this was explained in terms of the ability of cement to 'fix' a larger proportions of the chlorides to the surface. Bamforth [8.17] analysed a bulk of data and recommended the use of a C(o) value of 0.5% by weight of concrete. For a typical concrete of unit weight of about 2400 kg/m$^3$ this corresponds to 12 kg/m$^3$, which is slightly lower than the values recommended by Berke and Hicks [8.22].

### 8.3.4. Threshold Chloride Concentrations

The concepts of the threshold chloride concentrations were discussed in section 4.2.2. Usually, the threshold value is assigned to the concentration required at the steel level to cause initiation of corrosion, $C_i$, i.e. depassivation of the steel (point I in Fig. 4.2). However, extensive analysis of laboratory and field data suggested that the time at which the corrosion has propagated to such an extent that cracking and delamination of the concrete covered has started (point P in Fig. 4.2) can also be related to the build-up of a threshold chloride concentration, $C_p$. Values of $C_i$ and $C_p$ are reported in units of % wt. of cement or kg/m$^3$ of concrete. Values to be used for design purposes were provided in several references:

Bamforth [8.17]: $C_i = 0.4\%$, $C_p = 1.0\%$ by weight of cement;
Berke and Hicks [8.22]: $C_i = 0.9$ kg/m$^3$, $C_p = 3$ kg/m$^3$ of concrete.

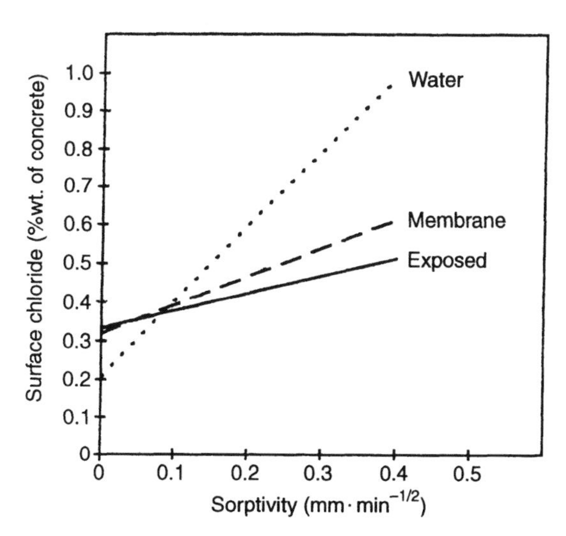


Fig. 8.7. Concentration of chlorides at the surface of the concrete as a function of the sorptivity of the concrete cover and the curing conditions. (After Ref. 8.17)

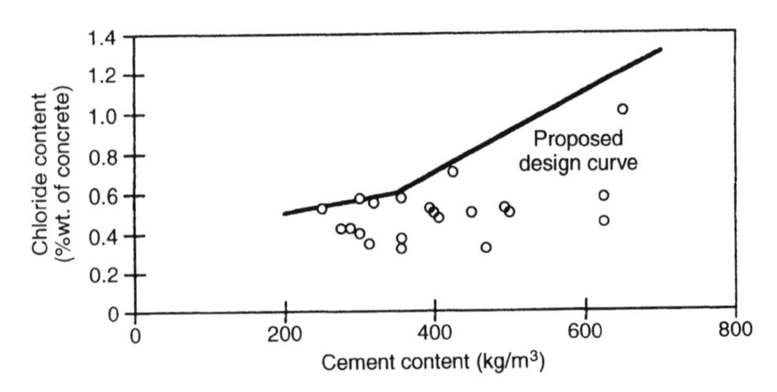


**Fig. 8.8.** Concentration of chlorides at the surface of the concrete as a function of the cement content in the mix. (After Ref. 8.17)

The two sets of values are practically the same; for a typical concrete containing 300 kg/m$^3$ of cement the percent values suggested by Bamforth would be translated into $C_i$ = 1.2 kg/m$^3$ and $C_p$ = 3.0 kg/m$^3$, which are similar to the values suggested by Berke and Hicks.

### 8.3.5. Service Life Calculations according to the Model

Calculations of service life, or the depth of cover and concrete quality required to achieve specified service life can be carried out using the concepts presented in Fig. 8.3 and the characteristic values provided in sections 8.3.2, 8.3.3 and 8.3.4.

Bamforth used this model and data to calculate the expected service life for concretes constructed according to the British Standards that are exposed to severe environments (Table 8.7). The calculated service life for onset of corrosion damage ($C_p$ = 1.0%) is usually less than 20 years, suggesting that the codes are not adequate for structures exposed to severe chloride environment which are expected to be in service for a long time.

Similar conclusions were reached by Berke who calculated how changes in $D_{eff}$ affect the chloride profiles at 50 years (Fig. 8.9) for severe chloride exposure conditions where C(o) = 17.8 kg/m$^3$. It is evident that decreasing $D_{eff}$ is an effective means of lowering chloride ingress. The benefits of increased concrete cover are also shown. However, in all the cases presented in this figure a concrete cover >100 mm is required. This is seldom practical, and costs, deadload or height considerations often dictate a smaller depth of cover.

Another approach in looking at the diffusion generated chloride profiles is to determine the chloride content as a function of time for a given depth. Figure 8.10 presents the predicted chloride content at 75 mm depth for two of the $D_{eff}$ values in Fig. 8.9 as a function of time. This is useful in determining the time to corrosion initiation when the design

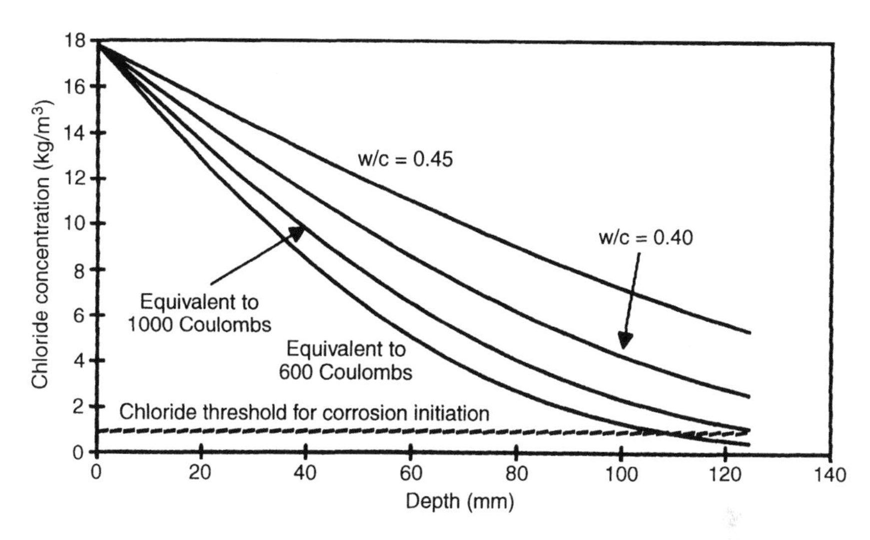


**Figure 8.9** Estimated chloride profiles for a marine wall at 18°C in the splash/tidal zone after 50 years assuming C(o) = 17.8kg/$m^3$ for the calculation. The concretes are of 0.45, 0.40 w/c ratio and high quality concretes with Coulomb values of 1000 and 600, with effective diffusion coefficients of $4.60 \times 10^{-12}$, $2.3 \times 10^{-12}$, $1.40 \times 10^{-12}$ and $0.99 \times 10^{-12}$ $m^2/s$, respectively.

calls for a fixed concrete cover. Note that the 0.45 w/c maximum design criteria ($D_{eff} = 4.6 \bullet 10^{-12}$ $m^2/s$) is not adequate for an 18°C exposure in a severe chloride environment.

The above calculations according to the model indicate that the current codes do not provide for sufficient service life when severe chloride environment is in question. The required quality of concrete (in terms of $D_{eff}$) and depth of concrete cover to achieve 50 and 120 years of service life in a severe chloride-containing environment were calculated by Bamforth [8.17] and are presented in Table 8.11. The calculations are for two criteria: corrosion initiation ($C_i$ = 0.4% by wt. of cement) as well as corrosion damage ($C_p$ = 1.0% by wt. of cement). It can be seen that to achieve the service life in question with concrete cover <50 mm or <75 mm, the values of $D_{eff}$ should be smaller than about $10^{-13}$ and $10^{-12}$ $m^2/s$, respectively. Such values can only be obtained with high-quality concretes containing mineral admixtures (Table 8.10 and Fig. 8.6). The $D_{eff}$ values for concretes with w/c ratio of 0.45 which are usually specified in the codes for severe chloride exposure will not meet the requirements in Table 8.11 for reasonable cover depth of 75 or 50 mm.

As can be seen above, the effects of reducing chloride ingress can be modelled quite readily. If chloride levels at the steel can be predicted, then increased time to corrosion initiation with calcium nitrite corrosion inhibitor can be modelled by setting higher values for the chloride threshold, as shown in Table 7.3.

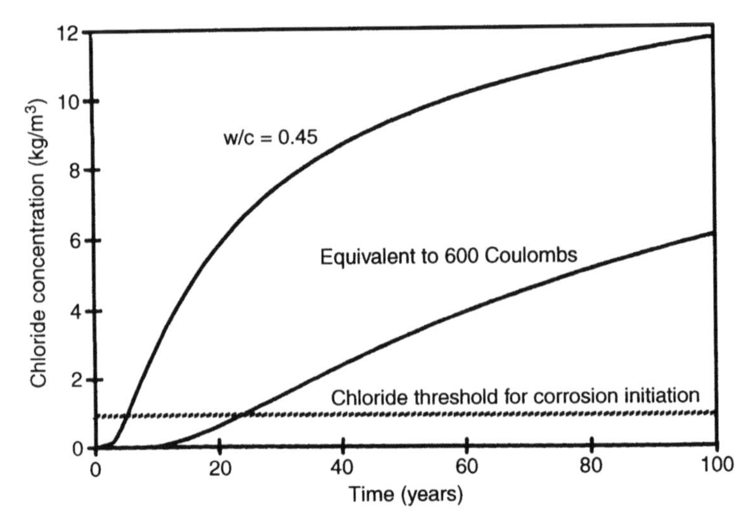


**Figure 8.10** Estimated chloride concentration for a marine wall at 18°C in hte splash/tidal zone at a depth of 75mm assuming C(o) = 17.8 kg/m³ for the calculation. The curves in the figure are for two of the concretes in Fig. 8.9 with w/c ratio of 0.45 and high quality concrete with 600 Coulomb value, having effective diffusion coefficients of $4.60 \times 10^{-12}$ and $0.99 \times 10^{-12}$ m²/s, respectively.

At this time, models for the extended life with epoxy-coated steel, galvanized steel, or cathodic inhibitors are not available; however, evidence showing that they may fail exists as noted Chapter 7. Estimates for epoxy-coated reinforcing bars and additions of inhibitor admixtures to the concrete are given in Chapter 10.

## APPENDIX 8.A SUGGESTED REVISIONS TO THE EUROPEAN PRESTANDARD EN 206: CONCRETE- PERFORMANCE, PRODUCTION, PLACING AND COMPLIANCE CRITERIA (FINAL DRAFT, APRIL 1997)

The current appendix summarizes suggestions for revisions of the European prestandard EN 206 to issues related to the quality of concrete required for control of the corrosion of steel in concrete. It is based on data provided in a draft in document prEN 206: Final draft, April 1997. It is presented here because it is a good example of a comprehensive approach for designing concrete for durability control. We believe that this the more rational approach which will eventually take over the existing codes. However, it should be emphasized that the document addressed here is a preliminary document and not the final revision.

The exposure classes according to this document are presented in Table 8.A.1. Exposure conditions are classified in terms of the mecha-

**Table 8.11.** Estimated Concrete Quality and Depth of Concrete Cover to Achieve 50 and 120 years of Service Life in a Severe Chloride Environment.[a]

| *Cover depth (mm)* | *Maximum effective diffusion coefficient ($m^2/s$)* | | | |
|---|---|---|---|---|
| | $C_i = 0.4\%$ | | $C_p = 1.0\%$ | |
| | $t = 50$ *yrs* | $t = 120$ *yrs* | $t = 50$ *yrs* | $t = 120$ *yrs* |
| 30 | $1.06\times10^{-13}$ | $4.42\times10^{-14}$ | $2.17\times10^{-13}$ | $9.06\times10^{-14}$ |
| 40 | $1.89\times10^{-13}$ | $7.86\times10^{-14}$ | $3.87\times10^{-13}$ | $1.61\times10^{-13}$ |
| 50 | $2.95\times10^{-13}$ | $1.23\times10^{-13}$ | $6.04\times10^{-13}$ | $2.52\times10^{-13}$ |
| 65 | $4.98\times10^{-13}$ | $2.07\times10^{-13}$ | $1.02\times10^{-12}$ | $4.25\times10^{-13}$ |
| 75 | $6.63\times10^{-13}$ | $2.76\times10^{-13}$ | $1.36\times10^{-12}$ | $5.66\times10^{-13}$ |
| 90 | $9.54\times10^{-13}$ | $3.98\times10^{-13}$ | $1.96\times10^{-12}$ | $8.16\times10^{-13}$ |
| 100 | $1.18\times10^{-12}$ | $4.91\times10^{-13}$ | $2.42\times10^{-12}$ | $1.01\times10^{-12}$ |

[a]After Ref. 8.17.

nisms that may cause deterioration rather than general environmental conditions which are specified in existing codes. Environmental conditions which may lead to corrosion of steel are subdivided into two classes, depending on the corrosion initiation mechanism: carbonation- and chloride-induced corrosion. Each of these is further divided into exposure classes which take into account the moisture conditions. In the case of the chloride-induced corrosion consideration is also given to the source of chlorides.

The requirements from the concrete for the different exposure classes are given in Table 8.A.2. The specifications include three requirements: maximum w/c ratio, minimum cement content and minimum strength class. The strength class includes two values: the first is the characteristic cube strength and the second is the characteristic cylinder strength. For a given system, the three parameters are related and a choice of two of them will control the third. The choice of combination is at the discretion of the producer. The choice to be made should be based on considerations of the concrete system. For example, it has been suggested that for concretes containing fly-ash or prepared with blended cements it is preferred to specify the concrete in terms of the minimum strength class rather than the maximum w/c ratio (see discussion in section 8.1.2 and reference [8.12]).

Special attention is given to the specification of the minimum cement content in concretes containing fly-ash and silica fume. The specifications are based on the concepts of the efficiency factor *k*, which is in the range of 0.2 to 0.4 for fly-ash (depending on the class of cement with which the fly-ash is combined) and 1.0 to 2.0 for silica fume (the higher value is recommended to be used when the specified w/c ratio is less than 0.45; the lower value is recommended to be used when the corrosion is induced by carbonation and the specified w/c ratio is greater than 0.45).

**Table 8A.1.** Exposure Classes. (According to prEN 206: Final draft, April 1997)

| *Deterioration mechanism* | *Exposure class* | *Description of the environment* | *Examples* |
|---|---|---|---|
| No risk of corrosion or attack | XO | Not aggressive to concrete | Concrete inside buildings with very low air humidity. |
| Corrosion induced by carbonation | XC1 | Dry | Concrete inside buildings with low air humidity |
| | XC2 | Wet, rarely dry | Parts of water retaining structures. Many foundations. |
| | XC3 | Moderate humidity | Concrete inside buildings with moderate or high air humidity. External concrete sheltered from rain. |
| | XC4 | Cyclic wet and dry | Surfaces subject to water contact, not within exposure class XC2. |
| Corrosion induced by chlorides from sea water | XS1 | Exposed to airborne salt but not in direct contact with sea water | Structures near to or on the coast. |
| | XS2 | Submerged | Parts of marine structures. |
| | XS3 | Tidal, splash and spray zones | Parts of marine structures. |
| Corrosion induced by chlorides | XD1 | Moderate humidity | Concrete surfaces exposed to direct spray containing chlorides |
| | XD2 | Wet, rarely dry | Swimming pools. Concrete exposed to industrial waste waters containing chlorides. |
| | XD3 | Cyclic wet and dry | Parts of bridges. Pavements. Car park slabs |

**Table 8A.1.** *continued*

| | | | |
|---|---|---|---|
| Freeze–thaw attack | XF1 | Moderate water saturation, without de-icing agent | Vertical concrete surfaces exposed to rain and freeezing |
| | XF2 | Moderate water saturation, with de-icing agent | Vertical concrete surfaces of road structures exposed to freezing and airborne de-icing agents |
| | XF3 | High water saturation, without de-icing agent | Horizontal concrete surfaces exposed to rain and freezing |
| | XF4 | High water saturation, with de-icing agent | Road and bridge decks exposed to de-icing agents and vertical concrete surfaces exposed to direct spray containing de-icing agents and freezing |
| Chemical attack[a] | | | |
| | XA1 | Slightly aggressive chemical environment | |
| | XA2 | Moderately aggressive chemical environment | |
| | XA3 | Highly aggressive chemical environment | |

[a] Classification of the concentration of the aggressive substances is given in the table below. The most onerous value for any single chemical determines the class.

**Table 8A.2.** Limiting Values for Composition and Properties of Concrete. (According to prEN206: Final draft, April 1997)

| | *Exposure classes* | | | | | | | | | | | | | | | | | |
|---|---|---|---|---|---|---|---|---|---|---|---|---|---|---|---|---|---|---|
| | *No risk of attack or corrosion* | *Carbonation-induced corrosion* | | | | *Chloride-induced corrosion* | | | | | | *Aggressive chemical environments* | | | *Freeze–thaw attack* | | | |
| | | | | | | *Sea water* | | | *Chloride other than sea water* | | | | | | | | | |
| | *XO* | *XC 1* | *XC 2* | *XC 3* | *XC 4* | *XS 1* | *XS 2* | *XS 3* | *XD 1* | *XD 2* | *XD3* | *XA 1* | *XA 2* | *XA 3* | *XF 1* | *XF 2* | *XF 3* | *XF 4* |
| Maximum w/c ratio[a] | – | 0.65 | 0.60 | 0.55 | 0.50 | 0.50 | 0.45 | 0.45 | 0.55 | 0.55 | 0.45 | 0.55 | 0.50 | 0.45 | 0.55 | 0.55 | 0.50 | 0.45 |
| Maximum strength class[b] | C12/15 | C20/25 | C25/30 | C30/37 | C30/37 | C30/37 | C35/45 | C35/45 | C30/37 | C35/45 | C35/45 | C30/37 | C30/37 | C35/45 | C30/37 | C25/30 | C30/37 | C30/37 |
| Minimum cement conten (kg/m³) | – | 260 | 280 | 280 | 300 | 300 | 320 | 340 | 300 | 300 | 320 | 300 | 320 | 360 | 300 | 300 | 320 | 340 |
| Minimum air content (%) | – | – | – | – | – | – | – | – | – | – | | – | – | – | – | 4.0[c] | 4.0[c] | 4.0[c] |
| Other requirements | | | | | | | | | | | | Sulphate- resistant cement[d] | | | Freeze–thaw-resistant aggregates | | | |

[a] From the limiting values for w/c ratio, strength class and cement content either the combination of w/c ratio and strength class or the combination of w/c ratio and cement content apply. The choice of the combination is at the discretion of the producer.
[b] Minimum strength classes were determined from the relationship between water/cement ratio and the strength class of concrete made with cement of strength class CE 32.5.
[c] If there is no use of entrained air then the performance of concrete has to be tested according to EN/XXX (Frost testing) in comparison with a concrete for which freeze–thaw resistance for the relevant exposure class is proven.
[d] Sulphate-resistant cement shall be used when $SO_4$ leads to exposure class XA2 or XA3.
If provision is made for two levels of sulphate-resistant cements in the standard, then either moderate- or high-sulphate-resisting cement shall be used for exposure class XA2 (and in exposure class XA1 when applicable) and high-sulphate-resistant cement shall be used for exposure class XA3.

For fly-ash additions the following guidelines are suggested:

- The maximum amount of fly-ash to be taken into account for the *k*-value concept shall meet the requirement fly-ash/cement $\leq 0.33$.
- If a greater amount of fly-ash is added, the excess shall not be taken into account for the calculation of the water/(cement + *k* • fly-ash) ratio, and the minimum cement content.
- The minimum cement content required for the exposure class may be reduced by a maximum amount of *k* • (minimum cement content – 200) kg/m$^3$ and additionally the amount of (cement + fly-ash) shall not be less than the minimum cement content required in Table 8.A.2.

For silica fume additions the following guidelines are suggested:

- The maximum amount of silica fume to be taken into account for the water/cement ratio and the cement content shall meet the requirement silica fume/cement $\leq 0.11$.
- If a greater amount of silica fume is added, the excess shall not be taken into account in the *k*-value concept.
- The amount of (cement + *k* • silica fume) shall not be less than the minimum cement content required in Table 8.A.2 for the relevant exposure class.
- The minimum cement content shall not be reduced by more than 30 kg/m$^3$ in concrete for use in exposure classes for which the minimum cement content is $\leq$ 300 kg/m$^3$.

## REFERENCES

8.1. European pre-standard ENV 206, Concrete- Performance, Production, Placing and Compliance Criteria, 1992.
8.2. Building code requirements for reinforced concrete (ACI 318-89) and commentary (ACI 318R-89). Manual of Concrete Practice, Part 3, American Concrete Institute, Detroit, 1994.
8.3. ACI Committee 201, Guide to durable concrete. Manual of Concrete Practice, Part 1, American Concrete Institute, Detroit, 1994.
8.4. ACI Committee 357, Guide for design and construction of fixed off-shore concrete structures. Manual of Concrete Practice, Part 4, The American Concrete Institute, Detroit, 1994.
8.5. ACI Committee 222, Corrosion of metals in concrete. Manual of Concrete Practice, Part 1, The American Concrete Institute, Detroit, 1994.
8.6. British Standard 8110: 1985, Structural use of concrete, part 1. Code of practice for design and construction.
8.7. ACI Committee 301, Specifications for structural concrete for building. Manual of Concrete Practice, Part 3, The American Concrete Institute, 1994.
8.8. CEB Deign Guide: Durable concrete structures, Comite Euro-International du Beton, Thomas Telford, UK, 1992.
8.9. Eurocode 2: Design of Concrete Structures, European pre-standard ENV 1992.

8.10. British Standard PD 6534:1993, Guide to use in the UK of DD ENV 206: 1992 Concrete: performance, production, placing and compliance criteria.
8.11. Barker, A.P. & Mathews, J.D., Concrete durability specification by water/cement or compressive strength for European cement types. In *Durability of Concrete*, ed. V.M. Malhotra. ACI SP-145, American Concrete Institute, Detroit, 1994, pp. 1135–59.
8.12. Browne, R.D., Geoghegan, M.P. & Baker, A.F., Analysis of structural condition from durability results. In *Corrosion of Reinforcement in Concrete Construction*, ed. A. P. Crane. Soc. Chem. Ind. UK, 1983, pp. 193–222.
8.13. Van Daveer, J.R., Techniques for evaluating reinforced concrete bridge decks. *J. Am. Concrete Inst.*, **72**(12) (1975), 697–704.
8.14. Browne, R.D. & Baker, A.F., The performance of structural concrete in a marine environment. In *Development in Concrete Technology*, ed. F.D. Lyden. Applied Science Publishers, London, 1979, pp. 111–49.
8.15. Browne, R.D., Low maintenance concrete specification versus practice? *Proc. 2nd Int. Conf. Maintenance of Marine Structures*, Thomas Telford, UK, 1986.
8.16. Marshall, A.L., *Marine Concrete*. Van Nostrand Reinhold, NY, 1990.
8.17. Bamforth, P.B., Specification and design of concrete for the protection of reinforcement in chloride-contaminated environments. Paper presented at 'UK Corrosion & Eurocorr 94', Bournemouth International Centre, 31 October–3 November, UK, 1994.
8.18. ACI Committee 224, Control of cracking in concrete structures. *Concrete International-Design and Construction*, **2**(10) (1980), 35–76.
8.19. Beeby, A.W., Concrete in the Oceans – Cracking and Corrosion. Tech. Rep. No. 2, CIRIA/UEG, Cement and Concrete Association, UK, 1979.
8.20a. Bentur, A. & Jaegermann, C., Effect of curing and composition on the development of properties in the outer skin of concrete. *ASCE J. Materials in Civil Engineering*, **3**(4) (1990), 252–62.
8.20b. Bentur, A. & Jaegermann, C., Effect of curing in hot environment on the properties of the concrete skin. Research Report, Building Research Station, Technion-Israel Institute of Technology, Haifa, Israel, 1989 (in Hebrew).
8.21. Parrott, P.J., Design for Avoiding Damage Due to Carbonation-Induced Corrosion. In *Durability of Concrete*, ed. V.M. Malhotra. ACI SP-145, American Concrete Institute, Detroit, 1994, pp. 283–98.
8.22. Berke, N.S. & Hicks, M.C., The life cycle of reinforced concrete decks and marine piles using laboratory diffusion and corrosion data. In *Corrosion Forms and Control for Infrastructure*, ed. V. Chaker. ASTM STP 1137, American Society for Testing and Materials, Philadelphia, 1992, pp. 207–31.

CHAPTER 9

# Repair and Rehabilitation

## 9.1. INTRODUCTION

The first step in the process of repair and rehabilitation is the assessment of the structure to evaluate the following:

- causes of the corrosion of the steel;
- the extent of the damage ;
- the expected progress of the damage;
- the influence of the damage on the structural safety of the structure, and on its serviceability.

On the basis of the initial assessment, the optimum strategy for repair and rehabilitation should be chosen. The strategy adopted should also take into account economic aspects, which include not only the cost of the repairs, but also the overall state of the structure and the additional time period it is expected to serve. Details and guidelines for the choice of strategies are provided in a RILEM document [9.1].

Four strategies can be generally considered:

(1) Replacement or reconstruction of the damaged components.
(2) Extensive repair to restore the integrity of the damaged components and to prevent additional corrosion.
(3) Local repair at regular intervals, and continuous monitoring of the development of damage.
(4) Provision of an alternative supporting structural system.

In the present chapter we will discuss mainly the strategies of extensive repair and of local repair.

## 9.2. FIELD ASSESSMENT TO QUANTIFY DAMAGE

Guidelines for assessment of corrosion damage have been issued by several organizations (e.g. CIRIA [9.2]). Assessment of the structure involves two stages. The first stage is a preliminary investigation to determine whether damage has occurred that requires repair. If it is concluded that repair is needed, a second stage consisting of an extensive investigation is launched. The results of this investigation should provide the information required for choosing the repair strategy and the details of the repair operation. Field test methods to evaluate the optimum conditions of the structure were reviewed in section 6.3.

An initial survey should include the following steps:

- Visual inspection and sounding tests (hammer-tapping and chain-drags) to determine damage: any rust stains, cracks and delamination should be noted. Crack width of any visible cracks should be recorded.
- Determination of cover thickness, of location of the steel, and of concrete quality. Non-destructive test methods should be used (equipment includes electromagnetic covermeter and Schmidt hammer).
- Determination of the depth of carbonation by phenolphthalein test of cover or of pieces of concrete broken from the surface.
- Measurement of the chloride content in cores or in pieces of concrete broken from the surface, or pieces of concrete drilled from the structure.

The more extensive survey which may follow if required should include the following steps:

- Mapping of potentials using half-cell techniques.
- Mapping areas of delamination and of damaged concrete, by means of ultrasonic testing or chain-drag.
- Determination of the electrical resistance of the concrete.
- Measuring surface properties of the concrete by surface absorption techniques or similar tests, and determining the strength of drilled concrete cores.
- Detailed study of depth of carbonation and of chloride ingress profiles by testing of drilled cores.
- Evaluating the nature of the cracks, and of their width and changes of width with time, to resolve whether they are 'live', i.e. cracks induced by thermal and loading effects, which are still active.
- Assessment of the structural conditions by full-scale loading tests in which the deflections under load are measured.
- Destructive tests to assess the reduction of the cross-section area of the steel.

## 9.3. PRINCIPLES OF REPAIR

Repair of concrete surfaces damaged by corrosion of steel consists of several steps (Fig. 9.1), and the repaired surface is composed of several elements as shown in Fig. 9.2. Both will be briefly considered here, and treated in greater detail in the following sections:

- Removal of cracked and delaminated concrete to expose all of the surface of the damaged steel.

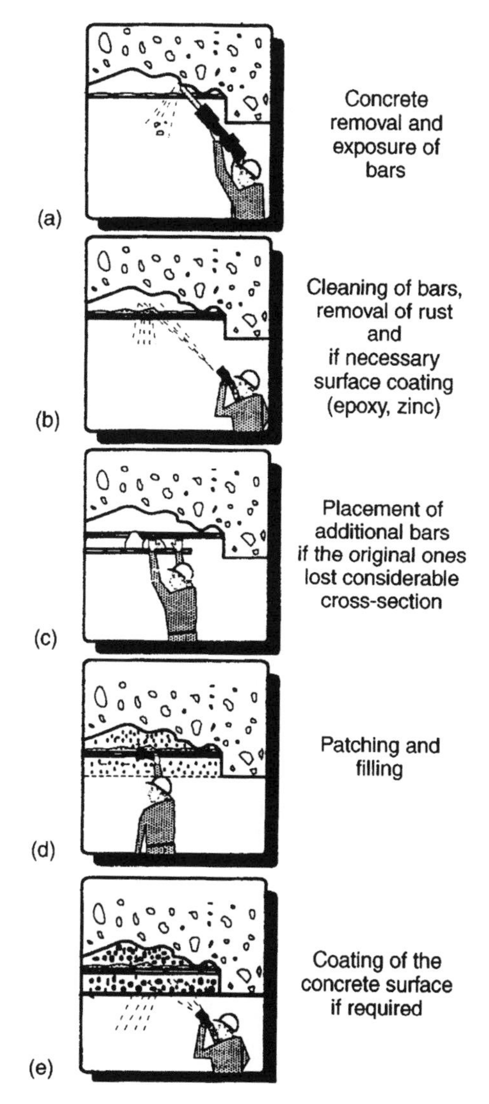


**Fig. 9.1.** Schematic description of the steps in repair operation of corroded steel in concrete. (A) Exposing of the corroded bar and removal of the damaged concrete around it. (B) Cleaning of the exposed bar to remove rust, and coating of the bar if necessary. (C) Replacement of the bar if necessary. (D) Priming of the concrete substrate (if necessary) and patching with repair material. (E) Application of external coating if necessary. (After Ref 9.6)

- Treatment of the steel to remove rusting layers and adding additional steel if the cross-section loss was heavy. Sometimes a protective coating is applied to the steel.

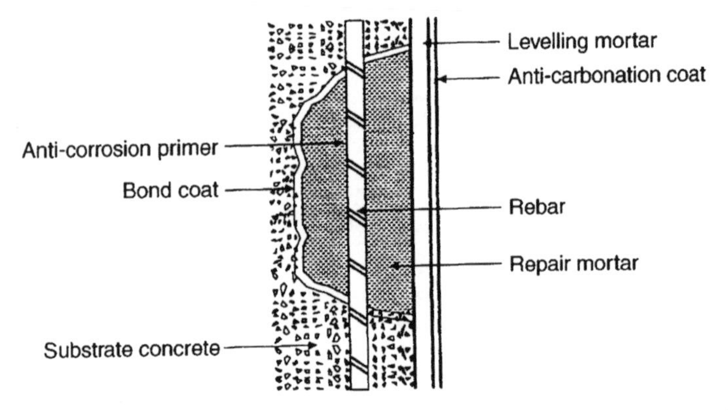


**Fig. 9.2.** Schematic description of the cross-section of repaired section. (Adapted from Ref. 9.4)

- Application of repair mortar or concrete to replace the damaged concrete that was removed. The repair mortar or concrete serves to protect the steel by both physical means (i.e. preventing ingress of deleterious substances) and by chemical means (providing repassivation). Repassivation can also be obtained by electrochemical methods. Sometimes a primer layer is applied at the interface between the old, sound concrete exposed around the steel, and the repair mortar or concrete.
- To enhance the protection provided by the mortar, or concrete, external membranes may be applied over the repaired section, or over all of the concrete surface.
- The repair mortar or concrete may also be applied in combination with the establishment of a cathodic protection system.

Not all of these means are required to be applied simultaneously. Repairs may not always include coating of the steel, or protection by external membranes. The choice of the combinations of treatments to be used, the materials to be used, and the application methods depend on a variety of factors. Most important are the cause of damage and the extent of the damage. An important guideline is that the repair system should restore the passivation of the steel that was disrupted either by carbonation or chlorides. In view of the differences between the two causes of depassivation, the repair systems chosen for carbonation- and chloride-induced corrosions are frequently different, as outlined below.

In the case of the carbonation-induced corrosion it is advisable to remove the carbonated concrete around the steel, and passivation is generated by the cement-based mortar repair layer. If the carbonation has propagated to a depth much greater than that of the concrete cover, it is not essential to remove all of the carbonated concrete. It was demonstrated [9.3] that in humid environments there can be sufficient diffusion of $OH^-$ ions into the carbonated layer from the alkaline repair mortar above it and from the non-carbonated concrete below it, to re-alkalize it.

The situation is usually more difficult in the case of chloride-induced corrosion. In may instances the depth of the penetration of chlorides considerably exceeds the depth of concrete cover. Removal of the concrete around the corroded steel bar, without removing the concrete below it, that is still contaminated with chlorides, will not assure repassivation, since the chloride ions will diffuse back into the repaired section. In such instances the conventional systems used for repair of carbonated-induced corrosion may not be adequate. An alternative which has been developed in recent years to combat this situation is chloride removal by electrochemical means. This will be discussed later in this chapter.

If for the reasons cited above it is not possible to apply the preferred strategy of restoration of the passivation of the steel, than it is essential to include in the repair system additional means to provide protection by other mechanisms:

- Applications of membranes and sealers on the repaired concrete surface to limit the moisture content of the concrete (section 7.4.3).
- Coating of the reinforcing steel bars.

In the following sections each of the components of the repair system will be discussed briefly. For more extensive treatment the reader is referred to references [9.4–9.6].

## 9.4. PATCH PREPARATION: REMOVAL OF DAMAGED CONCRETE AND CLEANING OF STEEL

The concrete around the corroded steel bar should be removed to a depth of at least 50 mm beyond the corroded portion. The bar should be exposed around all of the circumference of the corroded zone, with the clearance underneath the bar being at least 20 mm (Fig. 9.3). It is not necessary to completely expose reinforcing bars which do not show signs of corrosion. However, if in the process of patch preparation the concrete in contact with the non-corroded bars had been damaged to an extent that may detrimentally affect the concrete–steel bond, all of the reinforcing steel must be completely exposed. The repair mortar or concrete applied around them will restore the bond.

Hand-held pneumatic chisels can be used for concrete removal. When the concrete is very hard, more powerful means, such as high-pressure water jets, may have to be used.

In the removal of the concrete, care should be taken adequately to prepare the boundaries of the patch, to prevent feathered edge conditions, and obtain a patch geometry which minimizes the edge length.

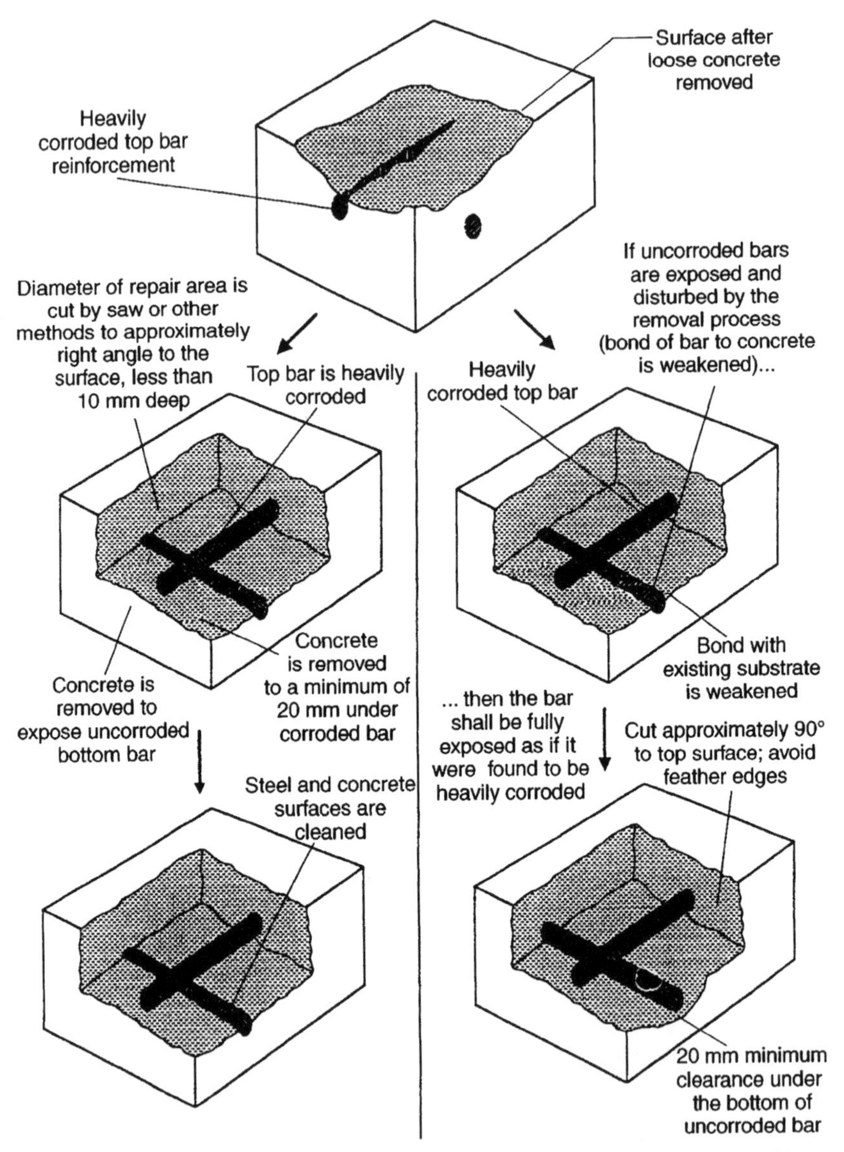


**Fig. 9.3.** Schematic description of the concrete removal around a corroded bar, and adjacent uncorroded bars as influenced by the bonding conditions of the uncorroded bar. On the left side, the uncorroded bottom bar is well bonded to the concrete. On the right side, the uncorroded bottom bar has been loosened from its grip to the concrete during the concrete removal. (After Ref 9.6)

This is usually achieved by disc cutting at right angles to the surface to a depth of 5–25 mm. Also, the patch boundaries should be set so as to minimize the length of the boundary between the existing concrete and the patch (Fig. 9.4).

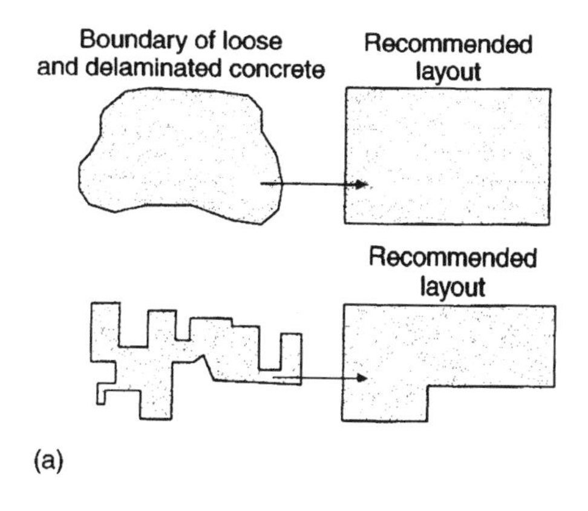

**Fig. 9.4.** Schematic description of the recommended layout of the patch exposed around the damaged areas. Sawing with a disc to obtain: (A) a perimeter of rectangular shape and (B) a patch wall at 90° to the surface. (After Ref. 9.6)

If heavy tools are being used, there is a need to inspect for damage that may have been developed around the excavated patch. Such damage may occur mainly in the form of fractured aggregates that have become loosened from the cementitious matrix. They should be removed with lighter tools.

The rust should be removed from the corroded steel surface, to allow intimate contact with the repair mortar or concrete so as to achieve repassivation. Grit blasting is a recommended method for this purpose, but there is often a reluctance to use this method because of environmental problems, such as the dust and noise generated during blasting. If high-pressure water jets are used to remove rust, it is necessary to introduce sand into the water jet, to enhance its abrasion efficiency. The use of wire brushes is not recommended, since these are not efficient in removal of rust.

After the completion of the removal of deteriorated concrete and rust, there is a need to clean the resulting surface from dust. The recommended method for this purpose is the use of industrial vacuum cleaners with small brush heads. If blasting with compressed air is used as a dust method, care should be taken to frequently service the filters for removal of oil and water.

## 9.5. TREATMENT OF THE STEEL

When the strategy of repair is based on repassivation of the steel by removal of all of the contaminated concrete and using a cementitious repair mortar or concrete, it may be argued that there is no need to treat the steel surface since it will be directly repassivated by the mortar or concrete applied in the patch. However, a risk exists that the compaction of this mortar/concrete may not be sufficiently effective. Thus, to ensure complete contact with a repassivation of all the steel, it is frequently recommended to treat the steel surface with alkaline slurry.

However, the application of such alkaline slurries (usually composed of Portland cement and water) may encounter difficulties as the slurry may thicken rapidly and not bond well to the repair mortar or concrete. The application of polymer–cement slurries, in which this difficulty is overcome, may raise other problems, such as reduced alkalinity or insulation of the steel from the passivating repair mortar. Thus, careful consideration should be given in specifying this kind of treatment.

When it is not possible to achieve repassivation (such as with chloride-contaminated concrete, as discussed in section 9.3) the repair strategy may be based on direct protection of the steel so as to seal and electrically isolate it from the surrounding repair mortar or concrete. Such a strategy is in marked contrast with the repassivation strategy discussed earlier. In the repassivation strategy isolation would be detrimental and therefore unwanted. Here, coating is an essential and vital component of the repair system. Epoxy resins formulated for this purpose are available.

However, there are several difficulties with this method: (i) it is not always possible to achieve complete coverage of the surface in the field application; (ii) there is a risk that due to the differences in the properties of the coated and uncoated portions of the bar, corrosion will develop in the uncoated part [9.7]; and (iii) if water can penetrate into the steel–epoxy interface, debonding followed by corrosion may occur [9.8].

In view of the limitations cited above, an alternative strategy for the protection of the steel may be offered, based on sacrificial cathodic protection using zinc coating, rather than isolation based on epoxy coating. The application of coating, whether epoxy or zinc, is a simple operation of paint brushing.

## 9.6. REPAIR MORTAR/CONCRETE

### 9.6.1. Materials Composition

If the area to be repaired is large and involves reconstruction of a large proportion of the cross-section, the preferred solution may be the use of concrete applied manually or by shotcreting. Proper shotcreting methods and mix design procedures are well documented and should be followed [9.9, 9.10]. However, many repair operations involve only patches, but require reconstruction of the surface of the concrete component. In such instances the repair material to be used is in the form of mortar. Repair mortars can be classified according to the composition of their binder as polymeric or cementitious mortars.

#### *9.6.1.1. Polymer Mortars*

Polymer mortars can be readily formulated to provide compositions with properties adjusted for efficient application. Adequate workability and flow properties are required, as well as controlled rapid setting. In addition, polymer mortars can provide strong bonding to the substrate, eliminating the need for a primer. They can provide an impermeable protective cover over the reinforcing bar. A variety of formulations are available with compressive strength values in the range of 50 to 100 MPa, and relatively high tensile strengths.

In spite of these favourable properties, such compositions have experienced performance problems due to the mismatch in properties between the polymer and the surrounding concrete. The polymer has a much higher coefficient of thermal expansion (typically about $65\times10^{-6}/°C$ for the polymer versus $12\times10^{-6}/°C$ for the concrete), and a much lower modulus of elasticity. In addition, the fast setting is frequently accompanied by considerable shrinkage. These conditions can lead to the build-up of internal stresses that may eventually cause cracking and delamination [9.11, 9.12]. Special formulations developed for concrete repair should minimize this incompatibility. This can be achieved by using a polymer binder with lower curing shrinkage, and proper grading of the aggregates in the mortar to minimize the polymer content in the mix. Test methods to evaluate incompatibility are available and should be used to characterize these mortars (e.g. ASTM C 884-92 Standard Test Method for Thermal Compatability Between Concrete and Epoxy-Resin Overlay).

The polymer binders normally used are formulated from either epoxy or polyester polymers. The polyester mortars are usually resin-rich. They can readily wet the substrate, and a primer is not required. Epoxy mortars are usually formulated without excess binder, and since they tend to wet less readily the concrete surface, a primer is often recommended.

#### *9.6.1.2. Cementitious Materials*

Polymeric mortars can provide physical protection to the steel, but do not provide chemical repassivation. Since repassivation is often the preferred strategy, and in view of some of the limitations of the polymer mortars associated with the mismatch in properties, there is a recent tendency to specify cementitious mortars for repair. However, traditional compositions consisting of cementitious binder and graded fine aggregate do not provide adequate performance. Therefore, modern cementitious mortars used for repair consist of more sophisticated formulations which often include combinations of cementitious components with polymeric additives.

The use of polymer latex either in the form of a liquid suspension or dry powder premixed in the mortar is common. Polymer latex used in cementitious mortars imparts improved characteristics over conventional mortars [9.13, 9.14] including:

- improved ductility and flexural strength;
- reduced permeability, but retaining the ability for vapour movement to take place in the set mortar;
- improved bond to the substrate;
- improved workability.

The polymer latex chosen for such formulations should be compatible with the cementitious matrix, so that the polymeric phase in the form of a film formed during curing will be stable in the alkaline matrix. Also, the content of air which is entrained by the surface active agents in the latex should not be excessive. These characteristics are addressed in ASTM C 1059-91 (Specification for Latex Agents for Bonding Fresh to Hardened Concrete).

Additional components in the formulations often include inorganic materials. Silica fume in combination with high-range water reducers (superplasticizers) can impart excellent flow properties, to repair mortars, as well as high strength and impermeability of the hardened material. Special inorganic compounds can be added to reduce shrinkage during setting and hardening by promoting reactions which lead to controlled expansion. These repair materials are known as shrinkage compensating repair formulations.

### 9.6.2. Primer Coating

Primer coating is required to assure bond between the existing old concrete and the new repair mortar or concrete when the repair mix can not readily wet the concrete substrate. The ability to wet is a function of the composition of the repair mix, and of the method of application. For example, in trowelling application of a relatively dry mix, a primer bond is necessary, whereas in sprayed techniques with adequate concrete mix composition this stage may not be needed.

The primer coat is usually a cement–water slurry, or a slurry in which some polymer latex is incorporated. It is simply brushed over the concrete surface. For the primer coat to be effective, it must be still in a 'tacky' state when the repair material is applied [9.15]. Drying and premature curing of the primer bond coat could lead to a layer which will cause separation rather than bonding. This has been observed to be the case when improper procedures were used, such as allowing too long an elapsed time between the application of the primer and that of the repair mortar, or the use of double primer coatings [9.16]. To prevent premature drying of the primer coat, it is essential to wet the concrete substrate and bring it to a saturated surface dry condition before applying the primer coat.

In some instances wetting of the substrate is undesirable; for example, when it is necessary to prevent movement of salts from the existing concrete into the repaired portion. An alternative for such cases is the use of an epoxy resin bond coat. Here too, the repair concrete or mortar must be applied while the epoxy coat is still wet and sticky. Mays [9.4] reported that the best bonding was obtained when the repair mortar or concrete hardened before the primer-bonding layer did. The time during which the epoxy primer maintains its workability is called 'open time' and it can be determined by standard tests.

### 9.6.3. Application Methods

Schematic descriptions of the various methods of application of repair mortar or concrete are provided in Fig. 9.5. The mix composition and properties should be adjusted for each of these methods. The choice of the application method depends on the size of the area needed to be repaired, and the nature of the repair, i.e. surface repair or full depth repair.

- **Dry packing** is adequate for repair of small areas, and can be used in locations where access to the repair is difficult.
- **Hand-applied trowelling** is a method used for small-scale local repairs. The mix should be non-sagging. Its application in repair of corroded steel is limited since it is difficult to apply it around exposed bars, and as a result the encapsulation of the bar is not effective.
- **Form and cast-in-place** repair is a common method. It requires high-quality formwork and mixes of sufficient fluidity. Since difficulties may be encountered in the compaction of the repair concrete, high fluidity mixes have been developed for this purpose.
- **Form-and-pump** repair is a relatively new technique which provides flexibility for the use of a variety of materials. It also requires high quality and stable formwork, as well as mixes which can be pumped readily. It can be used for vertical repairs as well as overhead repairs.

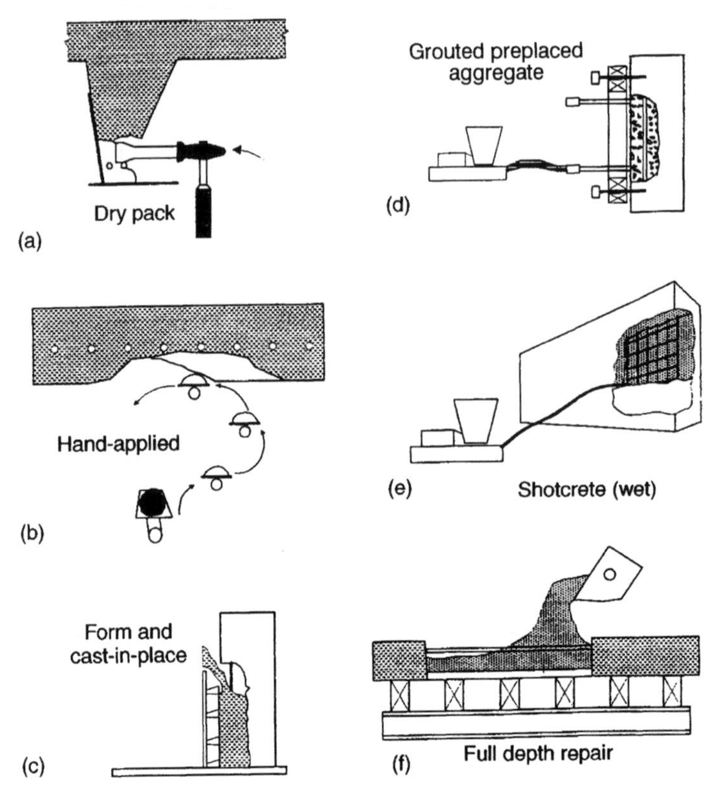


**Fig. 9.5.** Schematic description of the methods for application of repair mortar/concrete. (A) Dry packing. (B) Trowelling. (C) Form and cast in place. (D) Grouted preplaced aggregates. (E) Shotcreting. (F) Full depth repair. (After Ref. 9.6)

- **Grouted preplaced aggregate** repair techniques are sometimes used. The aggregates for such repairs are gap graded. They are placed into the cavity, and then a fluid grout is pumped in to fill the voids between the aggregates. An advantage of this system is the low shrinkage resulting from the point contact between the adjacent preplaced aggregates.
- **Shotcreting** is a technique which is suitable for repair of large areas. Wet and dry processes are available. Specifications for shotcreting equipment, mix compositions, and application procedures are available in several references (e.g. [9.9, 9.10]). High-quality concretes with excellent bond to the substrate can be achieved. A variety of sophisticated mix compositions are available, which may include various additives such as accelerators, silica fume and fibres.
- **Full depth repair** is used when reconstruction of a corroded member is required.

## 9.7. ELECTROCHEMICAL PROTECTION

The need to apply electrochemical protection as part of the repair strategy was discussed in section 9.3. This method is sometimes required in the case of chloride-induced corrosion, when the depth of chloride penetration is large, and it is not feasible to remove a large cross-section of the contaminated concrete. In such instances, electrochemical methods can be used to provide cathodic protection as outlined in section 7.4.5, or they can be used for chloride removal which will be discussed here briefly.

Electrochemical removal of chloride is a relatively new concept, and only limited field experience is yet available. A temporary anode (a steel mesh or other inert material) surrounded by electrolyte (usually a thick slurry of saturated lime solution or tap water) is placed over the concrete surface. A current density in the range of 0.5 to 2 $\mu A/m^2$ is impressed, and the chloride is induced to migrate towards the anode by the electrical field created between the temporary anode and the reinforcing steel bars. The amount of chloride removed is proportional to the charge passed between the anode and the reinforcing steel. Due to the uncertainties of this technique it has been suggested [9.1] that it should be applied to preventing corrosion in structures where the chloride has penetrated into the concrete cover, but the threshold concentration of chloride at the level of the steel has not been exceeded. It was also pointed out that this technique is cost-effective if applied before deterioration has become so severe as to require removal of the concrete cover or replacement of cracked and delaminated concrete.

Electrochemical means have been also suggested to restore alkalinity. In this case as well, a temporary anode is place on the concrete surface, but the solution surrounding it is a slurry of sodium carbonate. The impressed current leads to generation of $OH^-$ ions at the reinforcing steel surface, and to transport of the alkali ions from the electrolyte into the concrete. This migration leads to increase in the pH level of the pore solution around the reinforcing steel bars. The method is intended for application in cases of carbonation-induced corrosion. In view of the limited experience with the method to date, and other limitations, it has been suggested for use in structures where carbonation of the concrete cover has occurred, but has not yet been accompanied by corrosion damage [9.1].

## 9.8. SEALERS AND MEMBRANES

The need to apply sealing membranes as part of the repair strategy was highlighted in section 9.3. Sealing membranes are essential when: (i) it is not possible to repassivate the concrete (when the contaminating chloride can not be completely removed); (ii) when the depth of cover is so small (and can not be increased) that there is a risk of carbonation; or (iii) when penetration of chlorides to the steel level within a relatively short period may be expected, even if high-quality repair mortar or concrete is used.

The properties and application of membranes was discussed in section 7.4.3, and the principles highlighted there are applicable also when membranes are used as part of a repair system. Attention should be given to the area that needs to be covered by the membrane. It is preferable to specify that the membrane application covers all of the component (and perhaps the structure), not only the repaired section. This could avoid or delay corrosion that may be induced due to differences created between the steel in the repaired section and the steel beyond the repair boundaries.

## REFERENCES

9.1. RILEM Committee 124-SRC, Draft recommendation for repair strategies for concrete structures damaged by steel corrosion. *Materials and Structures*, **27**(171) (1994), 415–36.
9.2. Pullar-Strecker, P., *Corrosion Damaged Concrete: Assessment and Repair*. Butterworths-Construction Industry Research and Information Association, UK, 1987.
9.3. Biehr, T., Kropp, J. & Hilsdorf, H.K., Carbonation and realkalinization of concrete and hydrated cement paste. Vol. 3 in *Proc. 1st Int. RILEM Congress from Materials Science to Construction Materials Engineering*, ed. J.C. Mao. Chapman & Hall, 1987, pp. 927–34.
9.4. Mays, G., *Durability of Concrete Structures*. E & FN Spon, UK, 1992.
9.5. Allen, R.T.L., Edwards, S.C. & Shaw, J.D.N., *The Repair of Concrete Structures*. Blackie Academic & Professional, Chapman & Hall, UK, 1993.
9.6. Emmons, P.H., *Concrete Repair and Maintenance Illustrated*. R.S. Means Company, Inc., USA, 1994.
9.7. McCurrich, L.H., Cheriton, L.W. & Little, D.R. Repair systems for preventing further corrosion in damaged reinforced concrete. *Proc. 1st Int. Conf. on Deterioration of Reinforced Concrete in the Arabian Gulf*, CIRIA, UK, 1985.
9.8. Gledhill, R.A. & Kinloch, A.J., Environmental failure of structural adhesive joints. *Journal of Adhesion*, **6** (1974), 315–30.
9.9. ACI Committee 506, Proposed revision of: Specification for materials, proportioning and application of shotcrete. *Amer. Concrete Inst. Mater. J.*, **91**(1) (1994), 108–15.
9.10. Cooke, T.H., *Concrete Pumping and Spraying*. Thomas Telford Publications, UK, 1990.
9.11. Emberson, N.K. & Mays, G.C. The significance of property mismatch in the patch repair of structural concrete (1) Properties of repair systems. *Magazine of Concrete Research*, **42**(152) (1990), 147–60.
9.12. Emberson, N.K. and Mays, G.C. The significance of property mismatch in the patch repair of structural concrete (2) Axially loaded reinforced concrete members. *Magazine of Concrete Research*, **42**(152) (1990), 161–70.
9.13. ACI Committee 548, State of the Art Report on Polymer Modified Concrete. ACI Manual of Concrete Practice, American Concrete Institute, 1994.

9.14. Ohama, Y., *Handbook of Polymer-Modified Concrete and Mortars*. Noyes Publications, USA, 1995.
9.15. Judge, A.I. & Lambe, R.W., Selection and performance of substrate priming systems for cementitious repairs. In *Structural Faults and Repair '87*. Engineering Technics Press, Edinburgh, 1987, pp. 373–87.
9.16. Dixon, J.F. & Sunley, V.K., Use of bond coats in concrete repair. *Concrete*, **17**(8) (1983), 34–5.

CHAPTER 10

# Life-Cycle Cost Analysis

## 10.1. BASIC PRINCIPLES

### 10.1.1. Net Present Value

The cost of corrosion protection in a concrete structure is a combination of initial, maintenance, and repair costs occurring within the projected service life of the structure. In many cases the maintenance and repair costs can be considerably higher than the initial cost of a corrosion protection system. However, frequently only the initial costs are considered, and as a result the total costs may be higher.

The use of Net Present Value (NPV) analysis can be used to estimate the present cost of a future repair. For a repair that occurs in year $n$, the NPV is defined as:

$$\text{NPV} = \text{Cost} \times (1+D)^{-n} \tag{10.1}$$

where Cost is the price for the repair and $D$ is the discount rate or the interest rate less the inflation rate. For example, when the inflation rate is 4% and the interest rate is 8%, the discount rate, $D$, is 4% or 0.04.

The total life-cycle costs of a structure is the initial cost plus the sum of the NPV of all future costs. As the discount rate increases, the cost of future repairs is less significant. Thus, a lowering of the discount rate results in an increase in the NPV of future repairs. For the examples in this chapter the value of $D$ is chosen to be 4%, which is typical for several countries.

### 10.1.2. Time To Repair

The time for repair of a structure is related to the time for corrosion initiation and the time for corrosion damage to reach a point at which the concrete becomes distressed to the point where repairs are needed. This is shown schematically in Fig. 10.1 which is based upon a model by Weyers *et al.* [10.1], but was originally discussed by Tuutti [10.2]. In the treatment in this chapter it will be assumed that chloride-induced corrosion is the applicable depassivation mechanism. The same principles can be applied to carbonation-induced corrosion.

In Fig. 10.1 there is an initial damage which occurs during the construction stage, which is then repaired. Later on, chloride migration into the concrete is occurring for several years. When the chloride

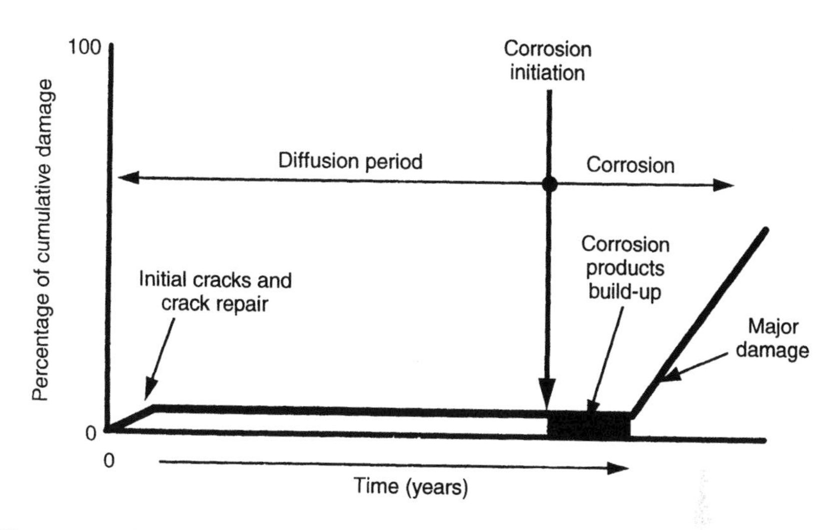


**Fig. 10.1.** Schematic description of the corrosion process based on concepts of Weyers (Ref. 10.1) and Tuuti (Ref. 10.2).

content reaches the critical level for depassivation to occur, pitting corrosion is initiated. Yet, it takes several more years for corrosion damage to cause concrete distress. For bridge decks in the United States this additional time is about 2–5 years [10.3]. Thus, the time to the first repairs in bridge decks is the depassivation (initiation) time plus about 3 years.

The above analysis shows that the time to repair is determined by the time to reach a critical chloride level for corrosion initiation and the corrosion rate after initiation. In Chapter 7, several corrosion protection mechanisms were discussed that could affect the time for chloride ingress, the critical chloride level for corrosion initiation, or the corrosion rate once corrosion was initiated.

The above analysis only holds for those cases where cracks do not play a significant role. In Chapter 8, additional design and specification information are given for crack control. Minimizing crack sizes, and the use of better quality concretes significantly reduces the effects of cracking as noted by several researchers [10.4–10.7].

In the following sections the above will be used in examples based upon bridge decks in a de-icing salt environment. The techniques are applicable to other structures with proper correction for environment and geometry factors and for typical repair costs.

## 10.2. SERVICE-LIFE COST EXAMPLE FOR A BRIDGE DECK IN A DE-ICING SALT ENVIRONMENT

In this section a typical bridge deck 200 mm thick with 50 mm of concrete cover is considered. A service life of 75 years is assumed. The base concrete is a mix of 0.4 w/c ratio. A typical average yearly temperature of 13°C will be assumed. Additional corrosion protection systems considered are two types of corrosion inhibitor, silica fume, epoxy-coated reinforcing bars, and some combination systems.

The additional typical initial costs (expressed as price in US dollars per square metre) are given in Table 10.1 for the corrosion protection systems considered. Note that these and all other costs should be adjusted based upon actual material used and local pricing. Repair costs are estimated to be $350/m^2$. It should be considered that only 10% of the deck requires repair, and therefore the average repair cost for the bridge deck is taken as $35/m^2$. A discount rate of 4% is used, and repairs are considered to be of adequate performance for 20 years (repairs within 5 years of the service life are not performed).

### 10.2.1. Time-to-Corrosion Initiation

Corrosion initiation is determined based upon diffusion of chloride into the concrete and upon the threshold chloride value for corrosion. The diffusion coefficients were assumed to be $1.63 \times 10^{-12}$ m²/s for the concrete at 0.4 w/c and $1.03 \times 10^{-12}$ m²/s for the concrete with silica fume. Figure 10.2 shows the increase in chloride content over time at the 50 mm cover depth. The pseudo surface chloride concentration was assumed to increase over time at a yearly rate of 0.6 kg/m³ until a constant value of 15 kg/m³ is reached. The calculations are based on numerical methods described in reference [10.7]. The results of these calculations will be presented here.

The times to reach critical chloride thresholds for corrosion initiation for the various corrosion protection systems can be calculated based on the principles presented in Fig. 10.2. The results are given in Table 10.2,

**Table 10.1.** Costs Per Square Metre for Corrosion Protection Systems in a Bridge Deck Used in Analysis.

| *Protection system* | *Costs ($/m²)* |
|---|---|
| No protection | 0 |
| Inhibitor 1, 10 l/m³ | 3.23 |
| Inhibitor 1, 15 l/m³ | 4.84 |
| Inhibitor 2, 5 l/m³ | 3.40 |
| Silica fume | 6.67 |
| Epoxy reinforcing bars | 3.77 |
| Silica fume + inhibitor 1, 10 l/m³ | 9.90 |
| Silica fume + inhibitor 1, 15 l/m³ | 11.52 |
| Silica fume + inhibitor 2, 5 l/m³ | 10.07 |
| Epoxy reinf. + inhibitor 1, 10 l/m³ | 7.00 |
| Epoxy reinf. + inhibitor 1, 15 l/m³ | 8.61 |
| Epoxy reinf. + inhibitor 2, 5 l/m³ | 7.17 |

along with the chloride threshold values. Note that there is no increase in chloride threshold levels for silica fume or epoxy-coated reinforcing bars.

### 10.2.2. Time-to-Damage after Corrosion Initiation

As noted earlier, the time to repair after initiation was chosen to be 3 years except in the case of inhibitor 2 which increases to 4 years, and for the epoxy-coated reinforcing bars which is 15 years. The effect of epoxy on reducing corrosion rates is predominant, therefore no additional improvement with inhibitor is given. The decrease in corrosion after initiation is generally accepted for epoxy-coated steels, but not well documented for other systems at this time.

The total time to first repair based upon initiation time and propagation time is given in Table 10.2. Also, times for additional repairs at 20-year intervals are presented in Table 10.2.

### 10.2.3. Service-Life Costs Analysis

Equation (10.1) is applied for the various repair times noted in Table 10.2. The NPV of each repair along with initial costs are given in Table 10.3, and the total costs are illustrated in the bar chart in Fig. 10.3.

Examination of Fig. 10.3 and Table 10.4 clearly shows the benefits of corrosion protection systems. In several cases the use of combination systems offer additional benefits as well as safety factors.

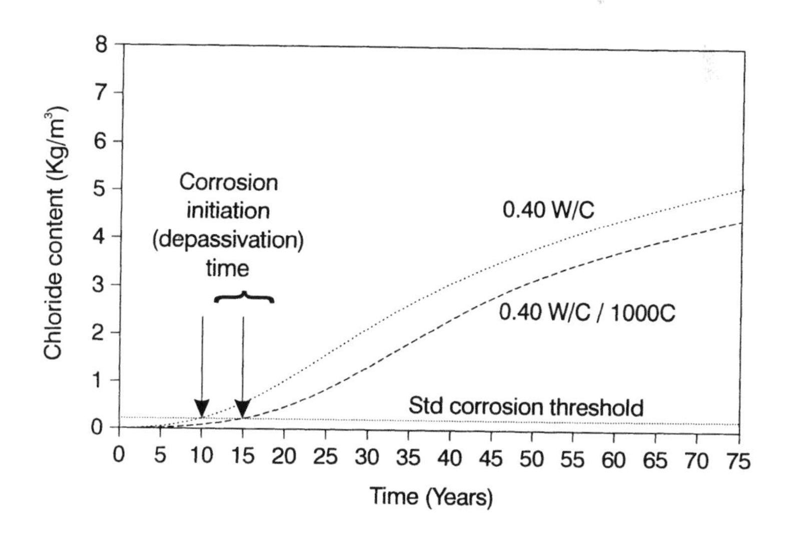


**Fig. 10.2.** Increase with time of chloride content at a depth of 50 mm in a 0.40 w/c ratio mix (0.40 w/c) and in a similar mix with silica fume having a Coulomb value below 1000 (0.40 w/c /1000 C) and schematic description of the determination of the corrosion initiation time.

**Table 10.2.** Times to Corrosion Initiation and Repairs for Various Corrosion Protection Systems (Years).

| *System* | *Chloride threshold (kg/m³)* | *Time to threshold* | *Time to Repair 1* | *Time to Repair 2* | *Time to Repair 3* |
|---|---|---|---|---|---|
| No protection | 0.9 | 17 | 20 | 40 | 60 |
| Inhibitor 1, 10 l/m³ | 3.6 | 37 | 40 | 60 | |
| Inhibitor 1, 15 l/m³ | 5.9 | 75+ | | | |
| Inhibitor 2, 5 l/m³ | 2.4 | 32 | 36 | 56 | |
| Silica fume | 0.9 | 22 | 25 | 45 | 65 |
| Epoxy reinforcing bars | 0.9 | 17 | 32 | 52 | |
| Silica fume + inhibitor 1, 10 l/m³ | 3.6 | 51 | 54 | | |
| Silica fume + inhibitor 1, 15 l/m³ | 5.9 | 75+ | | | |
| Silica fume + inhibitor 2, 5 l/m³ | 2.4 | 40 | 45 | 65 | |
| Epoxy reinf. + inhibitor 1, 10 l/m³ | 3.6 | 37 | 52 | | |
| Epoxy reinf. + inhibitor 1, 15 l/m³ | 5.9 | 75+ | | | |
| Epoxy reinf. + inhibitor 2, 5 l/m³ | 2.4 | 32 | 47 | 67 | |

In the above analysis cathodic protection was not considered. A typical cathodic protection system is \$43 to \$130/m². Even if the cost at the lower range (\$43/m²) is considered, this system is by far the most expensive one if used in new construction, even if the unlikely case is assumed that no

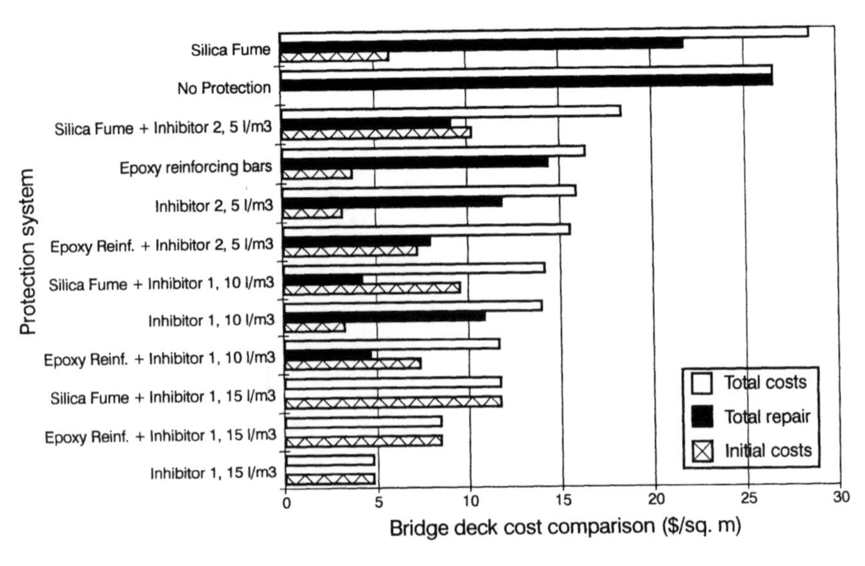


**Fig. 10.3.** Results of life-cycle cost analysis: initial, repair and total costs.

**Table 10.3.** Cost Analysis of Corrosion Protection Systems (\$/m$^2$).

| *System* | *Initial costs* | *NPV Repair 1* | *NPV Repair 2* | *NPV Repair 3* | *Total repair* | *Total costs* |
|---|---|---|---|---|---|---|
| No protection | 0 | 15.97 | 7.29 | 3.33 | 26.59 | 26.59 |
| Inhibitor 1, 10 l/m$^3$ | 3.23 | 7.29 | 3.33 | | 10.62 | 13.85 |
| Inhibitor 1, 15 l/m$^3$ | 4.84 | 0 | | | | 4.84 |
| Inhibitor 2, 5 l/m$^3$ | 3.40 | 8.53 | 3.89 | | 12.42 | 15.82 |
| Silica fume | 6.67 | 13.13 | 5.99 | | 21.86 | 28.53 |
| Epoxy reinforcing bars | 3.77 | 9.98 | 4.55 | | 14.53 | 18.30 |
| Silica fume + inhibitor 1, 10 l/m$^3$ | 9.90 | 4.21 | | | 4.21 | 14.11 |
| Silica fume + inhibitor 1, 15 l/m$^3$ | 11.52 | 0 | | | | 11.52 |
| Silica fume + inhibitor 2, 5 l/m$^3$ | 10.07 | 5.99 | 2.73 | | 8.73 | 18.80 |
| Epoxy reinf. + inhibitor 1, 10 l/m$^3$ | 7.00 | 4.55 | | | 4.55 | 11.55 |
| Epoxy reinf. + inhibitor 1, 15 l/m$^3$ | 8.61 | 0 | | | | 8.61 |
| Epoxy reinf. + inhibitor 2, 5 l/m$^3$ | 7.17 | 5.54 | 2.53 | | 8.07 | 15.24 |

**Table 10.4.** Initial Cost, Repair Cost and Total Cost of Different Systems and Protection Methods (\$/m$^2$).

| *System* | *Initial cost* | *Total repair cost* | *Total cost* |
|---|---|---|---|
| Silica fume | 6.67 | 21.86 | 28.53 |
| No protection | 0 | 26.59 | 26.59 |
| Silica fume + inhibitor 2, 5 l/m$^3$ | 10.07 | 8.73 | 18.8 |
| Epoxy reinforcing bars | 3.77 | 14.53 | 18.3 |
| Inhibitor 2, 5 l/m$^3$ | 3.40 | 12.42 | 15.82 |
| Epoxy reinf. + inhibitor 2, 5 l/m$^3$ | 7.17 | 8.07 | 15.24 |
| Silica fume + inhibitor 1, 10 l/m$^3$ | 9.90 | 4.21 | 14.11 |
| Inhibitor 1, 10 l/m$^3$ | 3.23 | 10.62 | 13.85 |
| Epoxy reinf. + inhibitor 1, 10 l/m$^3$ | 7.00 | 4.55 | 11.55 |
| Silica fume + inhibitor 1, 15 l/m$^3$ | 11.52 | 0 | 11.52 |
| Epoxy reinf. + inhibitor 1, 15 l/m$^3$ | 8.61 | 0 | 8.61 |
| Inhibitor 1, 15 l/m$^3$ | 4.84 | 0 | 4.84 |

components will need to be replaced over 75 years. If applied at 20 years it would have a NPV of \$37.88 based on an \$80/m$^2$ cost and no operating, maintenance or repair costs for the next 50 years.

The above analysis shows that the total service life cost of a concrete structure is not solely dependent on initial costs and that some, but not all, corrosion protection systems could significantly reduce total costs.

## REFERENCES

10.1. Weyers, R.E., Fitch, M.G., Larsen, E.P., Al-Qadi, I.L, Chamberlin, W.P. & Hoffman, P.C., Concrete Bridge Protection and Rehabilitation: Chemical and Physical Techniques. SHRP-S-668, Strategic Highway Research Program, National Research Council, Washington, DC, 1994.

10.2. Tuutti, K., *Corrosion of Steel in Concrete.* Swedish Cement and Concrete Research Institute, S-100 44 Stockholm, 1982.

10.3. Raupach, M., Corrosion of Steel in the Area of Cracks in Concrete-Laboratory Tests and Calculations Using a Transmission Line Model. In *Corrosion of Reinforcement in Concrete Construction*, Special Publication No. 183, eds C.L. Page, P.B. Bamforth & J.W. Figg. The Royal Society of Chemistry, Cambridge, UK, 1996.

10.4. Ohno, Y., Praparntanatorn, S. & Suzuki, K., Influence of Cracking and Water Cement Ratio on Macrocell Corrosion of Steel in Concrete. In Corrosion of Reinforcement in Concrete Construction, Special Publication No. 183, eds C.L. Page, P.B. Bamforth & J.W. Figg. The Royal Society of Chemistry, Cambridge, UK, 1996.

10.5. Beeby, A.W., Cracking, Cover, and Corrosion of Reinforcement. *Concrete International*, **5**(2) (1983), 35–40.

10.6. Berke, N.S., Dallaire, M.C., Hicks, M.C. & Hooper, R.J., Corrosion of Steel in Cracked Concrete. *Corrosion*, **49**(11) (1993), 934–43.

10.7. Berke, N.S., Dallaire, M.C., Hicks, M.C. & McDonald, A.C., Holistic Approach to Durability of Steel Reinforced Concrete. In *Concrete in the Service of Mankind: Radical Concrete Technology*, eds R.K. Dhir and P.C. Hewelet. E & FN SPON, UK, 1996.

# Index